H. C. Yarrow

**Report Upon Ornithological Specimens Collected in the Years 1871,**

**1872, and 1873**

H. C. Yarrow

**Report Upon Ornithological Specimens Collected in the Years 1871, 1872, and 1873**

ISBN/EAN: 9783337218720

Printed in Europe, USA, Canada, Australia, Japan

Cover: Foto ©berggeist007 / pixelio.de

More available books at **www.hansebooks.com**

ENGINEER DEPARTMENT, U. S. ARMY.

GEOGRAPHICAL AND GEOLOGICAL EXPLORATIONS AND SURVEYS
WEST OF THE ONE HUNDREDTH MERIDIAN.

First Lieut. GEO. M. WHEELER, Corps of Engineers, in charge.

# REPORT

UPON

# ORNITHOLOGICAL SPECIMENS

COLLECTED IN

## THE YEARS 1871, 1872, and 1873.

OFFICE OF THE CHIEF OF ENGINEERS,
*Washington, D. C., June 27, 1874.*
SIR: Lieut. George M. Wheeler has sent to this office reports by Dr. H. C. Yarrow on the ornithological specimens collected on the expeditions of the former in 1871, 1872, and 1873.

I have respectfully to request that they be printed at the Government Printing-Office, and that 1,500 copies be furnished on requisition from this office.

Very respectfully, your obedient servant,
A. A. HUMPHREYS,
*Brigadier-General and Chief of Engineers.*
Hon. WM. W. BELKNAP,
*Secretary of War.*

Approved by the Secretary of War:
H. T. CROSBY,
*Chief Clerk.*
JULY 1, 1874.

# TABLE OF CONTENTS.

UNITED STATES ENGINEER OFFICE,
EXPLORATIONS AND SURVEYS WEST OF THE 100TH MERIDIAN,
*Washington, D. C., June 8, 1874.*
SIR: I have the honor to submit, with the letter of Dr. H. C. Yarrow, the inclosed reports upon the ornithological specimens collected by this expedition in 1871, 1872, and 1873, and, in view of the reasons urged by him, suggest the propriety of their publication.

Very respectfully, your obedient servant,

GEO. M. WHEELER,
*Lieutenant of Engineers, in charge.*

Brig. Gen. A. A. HUMPHREYS,
*Chief of Engineers.*

UNITED STATES ENGINEER OFFICE,
EXPLORATIONS AND SURVEYS WEST OF THE 100TH MERIDIAN,
*Washington, D. C., June 2, 1874.*
SIR: I have the honor to submit the inclosed reports herewith, and respectfully suggest that it is of the utmost importance that they should be published at the earliest practicable moment, in order that this expedition may receive its due share of credit for priority of discoveries therein contained. I would propose at first to issue them each in pamphlet, octavo, which may be readily distributed and used (should occasion require) in compiling later the quarto volume on natural history.

Very respectfully, your obedient servant,

H. C. YARROW,
*Surgeon and Naturalist to the Expedition.*

Lieut. GEO. M. WHEELER,
*Corps of Engineers.*

UNITED STATES ENGINEER OFFICE,
EXPLORATIONS AND SURVEYS WEST OF THE 100TH MERIDIAN,
*Washington, D. C., October 3, 1874.*

The proofs of these reports, while passing through the press during the absence of the authors in the field, were corrected by Mr. Robert Ridgway, of the Smithsonian Institution, to whom thanks are due for his kindness in this regard, and who revised the nomenclature to correspond with the latest information upon the subject.

H. C. Y.

# REPORT UPON AND LIST OF BIRDS COLLECTED BY THE EXPEDITION FOR EXPLORATIONS WEST OF THE ONE HUNDREDTH MERIDIAN IN 1872; LIEUT. GEO. M. WHEELER, CORPS OF ENGINEERS, IN CHARGE.

By Dr. H. C. YARROW and HENRY W. HENSHAW,

FEBRUARY 15, 1873.

The following report upon the birds of Utah and Nevada, collected by the expedition under Lieutenant Wheeler, is based upon the specimens taken and observed during the months of July, August, September, October, November, and December, 1872.

While every possible facility was afforded by the commanding officer, it is much to be regretted that more time could not have been spent in a closer examination of the habits of the various birds seen, and, as a consequence, the notes must necessarily appear somewhat meager.

It should be taken into consideration that many of the lines of travel were over an almost arid waste of sand and sage-brush, with little vegetation and less water. Under such circumstances we might expect to find a great paucity of species; but with the exception of those varieties inhabiting the wooded localities, our collection will be found to represent a fair proportion of the western forms, our list numbering no less than one hundred and sixty-five different species.

It is also greatly to be regretted that, owing to the lateness of the season at which the field-work commenced, very few observations as to the nesting-habits of the birds could be taken, although quite a number of eggs were secured. It is hoped that the discovery of several species new to the fauna of Utah may prove of some value to our present ornithological knowledge.

A list of species taken during the expedition of 1871 is appended, together with a complete list of the birds of Utah as far as known, with annotations by Mr. Henshaw, assistant naturalist.*

The classification of the species noted is substantially the same as that of Professor Lilljeborg, of Upsala, (vide Proceedings Zoölogical Society of London, January, 1866,) and adopted provisionally by the Smithsonian Institution, and the nomenclature that adopted by Dr. E. Coues in his Check-List of North American Birds.

## TURDIDÆ.

*Turdus migratorius*, L.—Robin.

Usually found in the neighborhood of settlements, building close to houses, and exhibiting the same sociability as in the East. Very common at Provo, where a few years since it was unknown. Flocks seen in Beaver Cañon in September, and occasional ones in the cañons in December; it undoubtedly winters there in considerable numbers.

* This list appeared as a separate paper, and was printed in the Annals of the Lyceum of Natural History of New York, vol. xi, June, 1874.

*Galeoscoptes carolinensis*, (L.)—Catbird.
Common in settlements, frequenting the thickets. Habits and notes same in the East.

*Oreoscoptes montanus*, (Towns.)—Mountain Mocking-bird.
Common throughout Utah and Eastern Nevada. Generally found near settlements and not in mountains; rather shy and difficult to approach.

*Harporhynchus crissalis* (?), Henry.—Red-vented Thrush.
When within a few miles of Saint George, the southernmost settlement of Utah, a small flock of curved billed thrushes was observed, supposed to be of the species above named, but, owing to their excessive shyness, no specimens were secured. From the fact that Dr. E. Palmer found these birds breeding at Saint George, there seems little doubt as to the identification.

*Habitat.*—Southern Utah, valley of the Colorado, and Upper Rio Grande.

| No. | Name. | Sex. | Locality. | Date. | Collector. |
|---|---|---|---|---|---|
| 1a | Turdus migratorius ........ | ♂ | Beaver, Utah.......... | Sept. 22 | H. & Y. |
| A1 | ...... do .................. | ...... | Alcoholic ........ .... | ......... | H. & Y. |
| 49 | Mimus carolinensis.......... | ♂ jun. | Provo, Utah ...... .... | July 25 | H. & Y. |
| 141 | ...... do .................. | ♀ ad. | ....do .............. | Aug. 1 | H. & Y. |
| 157 | Oreoscoptes montanus...... | ♂ jun. | Fairfield, Utah ........ | Aug. 1 | Y. |
| 114 | ...... do .................. | ♀ jun. | Fountain Green, Utah. | Aug. 20 | H. |
| 413 | ...... do .................. | ♂ jun. | ....do .............. | Aug. 20 | H. |
| 143 | ...... do .................. | ♂ ad. | Salina, Utah.......... | Sept. 5 | H. |
| 156 | ...... do .................. | ♀ ad. | Fairfield, Utah........ | Aug. 1 | Y. |
| 171 | ...... do .................. | ♂ ad. | Panquitch, Utah...... | Sept. 17 | H. |
| 313 | ...... do .................. | ♂ ad. | Toquerville, Utah..... | Oct. 16 | H. & Y. |
| A2 | ...... do .................. | ...... | Alcoholic .... .... .... | ......... | H. & Y. |

CINCLIDÆ.

*Cinclus mexicanus*, Sw.—Water-Ouzel.
Numerous in the Provo River. One specimen taken on the Beaver River and one at Fillmore. Not observed elsewhere. It exhibits little shyness, permitting the close approach of a person, who may watch its interesting movements at leisure. Frequents exclusively the vicinity of rapids and falls, where it is to be seen constantly in motion, flying from rock to rock, and wading into the shallows, searching nervously for crustacea and water-insects, which form its food. The movements of its body are very peculiar, and consist of an emphatic, grotesque, downward jerk, constantly repeated, reminding one of a similar motion peculiar to some of the wrens, more particularly the Rock-Wren (*Salpinctes obsoletus*).

## SAXICOLIDÆ.

*Sialia arctica*, Sw.—Arctic Bluebird.

A small flock of migrants, first noticed in Eastern Nevada early in August; another near Gunnison, Utah, early in September. From this time until November 15 they were usually noticed in small detached companies pursuing their way southward. Frequently seen hovering in the air catching insects.

*Sialia mexicana*, Sw.—Western Bluebird.

But a single specimen seen at Fish Springs, Utah.

| No. | Name. | Sex. | Locality. | Date. | | Collector. |
|---|---|---|---|---|---|---|
| 150 | Sialia artica | ♂ ad. | Gunnison, Utah | Sept. | 8 | H. |
| 151 | ...... do | ♀ ad. | ....do | Sept. | 8 | H. |
| 265 | .... . do | ♂ ad. | Iron City, Utah | Oct. | 8 | H. |
| 387 | ...... do | ♀ | Beaver, Utah | Nov. | 10 | Y. & H. |

## SYLVIIDÆ.

*Regulus calendula*, (L.)—Ruby-crowned Wren.

A few individuals seen in Middle and Southern Utah in fall, usually accompanying flocks of the Titmice (*Paridæ*).

| No. | Name. | Sex. | Locality. | Date. | Collector. |
|---|---|---|---|---|---|
| 304 | Regulus calendula | ♀ ad. | North Creek, Utah.... | Sept. 26 | H. & Y. |

## PARIDÆ.

*Parus atricapillus*, L., var. *septentrionalis*, Harris.—Long-tailed Chickadee.

Common in cottonwood-groves near Provo River in July and November; not seen elsewhere. Habits and notes similar to eastern variety.

*Parus montanus*, Gamb.—Mountain Chickadee.

Said to be very common in mountains of Utah, but during entire season only three individuals were perceived and captured at Fillmore.

*Lophophanes inornatus*, (Gamb.)—Gray Titmouse.

Numerous in scrubby cedars near Iron City and Beaver. Seldom seen in companies of more than two or three, and more often singly.

*Psaltriparus plumbeus*, Bd.—Lead-colored Tit.

Found in same localities as above, but always in large flocks.

| No. | Name. | Sex. | Locality. | Date. | Collector. |
|---|---|---|---|---|---|
| 26 | Parus, var. septentrionalis .. | ♀ jun. | Provo, Utah............ | July 1 | H. |
| 24 | ...... do .................... | ♂ ad. | ....do ................. | Aug. 3 | H. |
| 25 | ...... do .................... | ♀ ad. | ....do ................. | Aug. 3 | H. |
| 142 | ...... do .................... | ♂ jun. | ....do ................. | Aug. 3 | H. |
| 436 | ...... do .................... | ♂ | ....do ................. | Nov. 25 | H. & Y. |
| 414 | Parus montanus............. | ♀ | Fillmore, Utah........ | Nov. 17 | H. & Y. |
| 415 | ...... do .................... | ♀ | ....do ................. | Nov. 17 | H. & Y. |
| 417 | ...... do .................... | ♂ | ....do ................. | Nov. 17 | H. & Y. |
| A3 | ...... do .................... |  | Alcoholic .... ..... .... | ......... | H. & Y. |
| A4 | ...... do .................... | ...... | ....do .... .... | ......... | H. & Y. |
| 247 | Lophophanes inornatus..... | ♂ | Iron City, Utah........ | Oct. 5 | H. |
| 258 | ...... do .......... | ♀ | ....do ................. | Oct. 6 | H. |
| 261 | ...... do .................... | ♀ ad. | ....do ................. | Oct. 8 | H. |
| 262 | ...... do .................... | ♂ | ....do ................. | Oct. 8 | H. |
| 263 | ...... do .................... | ♂ ad. | ....do ................. | Oct. 8 | H. |
| 264 | ...... do .................... | ♀ | ....do ................. | Oct. 8 | H. |
| 390 | ...... do .................... | ♂ | Beaver, Utah ......... | Nov. 11 | Y. & H. |
| 391 | ...... do .................... | ♀ | ....do ................. | Nov. 11 | Y. & H. |
| 392 | ...... do .................... | ♂ | ....do ................. | Nov. 11 | Y. & H. |
| 393 | ...... do .................... | ♂ | ....do ................. | Nov. 11 | Y. & H. |
| 394 | ...... do .................... | ♀ | ....do ................. | Nov. 11 | Y. & H. |
| 395 | ...... do .................... | ♀ | ....do ................. | Nov. 11 | Y. & H. |
| 396 | ...... do .................... | ♀ | ....do ................. | Nov. 11 | Y. & H. |
| 397 | ...... do .................... | ♂ | ....do ..... ...... ..... | Nov. 11 | Y. & H. |
| 403 | ...... do .................... | ♂ | Cove Creek, Utah..... | Nov. 13 | Y. & H. |
| 238 | Psaltriparus plumbeus...... | ♂ | Iron City, Utah........ | Oct. 5 | H. |
| 239 | ...... do .................... | ♂ | ....do ................. | Oct. 5 | H. |
| 240 | ...... do .................... | ♀ | ....do ................. | Oct. 5 | H. |
| 241 | ...... do ...... | ♂ | ....do ................. | Oct. 5 | H. |
| 242 | ...... do .................... | ♂ | ....do ................. | Oct. 5 | H. |
| 243 | ...... do ...... .:......... | ♂ | ....do ................. | Oct. 5 | H. |
| 244 | ...... do .................... | ♀ | ....do ................. | Oct. 5 | H. |
| 245 | ...... do .................... | ♂ | ....do ................. | Oct. 5 | H. |
| 246 | ...... do .................... | ♀ | ....do ................. | Oct. 5 | H. |
| 386 | ...... do .................... | ♀ | Beaver, Utah ......... | Nov. 10 | H. |
| A5 | ...... do .................... | ...... | Alcoholic ............. | ......... | H. & Y. |
| A6 | ...... do .................... | ...... | ....do ................. | ......... | H. & Y. |

SITTIDÆ.

*Sitta carolinensis*, Gm., var. *aculeata*, Cass.—Slender-billed Nuthatch.

*Sitta pygmœa*, Vig.—California Nuthatch.

Both species seen upon a single occasion in the heavy pine-timber of the Wahsatch Mountains.

TROGLODYTIDÆ.

*Troglodytes œdon*, V., var. *parkmanni*, Aud.—Western House-Wren.

Seldom seen and only in mountains; frequents brush-heaps and thick undergrowth. Habits about same as eastern variety (*œdon*).

*Salpinctes obsoletes*, (Say.)—Rock-Wren.

First seen at Gunnison, Utah, in September; and from this point southward a gradual increase in numbers was noted, until at Toquerville and Saint George hundreds were seen, chirping cheerily and enlivening the bleak and desolate volcanic rocks with their agile movements.

*Catherpes mexicanus*, (Sw.), var. *conspersus*, Ridg.—White-throated Wren.

Apparently rare at Toquerville, at which place a few specimens were secured. Like the preceding species, it inhabits the volcanic rocks, and the agility and celerity of its movements, as it springs from point to point, uttering its shrill and piercing note, is truly remarkable.

*Telmatodytes palustris*, (Wils.), var. *paludicola*, Bd.—Western Long-billed Marsh-Wren.

During the latter part of the month of July these interesting little birds were to be seen by thousands in the marshes near the Provo River, and their nests, carefully built and supported in the rushes, were perceived on all sides.

*Cistothorus stellaris*, (Licht.)—Short-billed Marsh-Wren.

This bird is somewhat rare in the Eastern United States; and its western limit, as far as known, being the Loup Fork of the Platte. While at Provo we received undoubted evidence of its existence in the marshes of the river, where it lived in company with the preceding species. Although no individuals were actually captured, the nests and eggs were seen, which had been secured in this locality.

*Thryothorus bewickii*, (Aud.), var. *leucogaster*, Gould.—White-bellied Wren.

Individuals occasionally seen in the southern part of Utah, and seemingly equally at home in the sparse shrubbery of the mountain-sides and in the valleys.

*Campylorhynchus brunneicapillus*, (Lafr.)—Cactus-Wren.

A single individual of this species was captured a few miles north of Saint George in October; two others being seen at the same time. It is believed that this is the most northern locality in which this bird has been taken. Although not chronicled from Arizona, there seems to be no doubt but that it is somewhat abundant there. One specimen was secured in 1871 by Lieutenant Wheeler's party, and others were seen.

| No. | Name. | Sex. | Locality. | Date. | Collector. |
|---|---|---|---|---|---|
| 103 | Troglodytes ædon, var. parkmanni. | ♀ jun. | Wahsatch Mountains, Utah. | Aug. 17 | H. |
| 164 | ...... do ............... | ♂ ad. | Otter Creek, Utah .... | Sept. 14 | H. |
| 141 | Salpinctes obsoletus........ | ♀ | Gunnison, Utah........ | Sept. 5 | H. |
| 245 | ...... do ............... | ♀ | Toquerville, Utah..... | Oct. 13 | H. |
| 248 | ...... do ............... | ♂ | ....do ............... | Oct. 13 | H. |
| 289 | ...... do ............... | ♂ | ....do ............... | Oct. 13 | H. |
| 290 | ...... do ............... | ♂ | ....do ............... | Oct. 13 | H. |
| 291 | ...... do ............... | ♀ | ....do ............... | Oct. 13 | H. |
| 299 | ...... do ............... | ♀ | ....do ............... | Oct. 13 | H. |
| 300 | ...... do ............... | ♂ | ....do ............... | Oct. 13 | H. |
| 305 | ...... do ............... | ♀ | ....do ............... | Oct. 13 | H. |
| 298 | ...... do ............... | ♂ | ....do ............... | Oct. 14 | H. |
| 311 | ...... do ............... | ♀ | ....do ............... | Oct. 16 | Y. & H. |
| 312 | ...... do ............... | ♀ | ....do ............... | Oct. 16 | Y. & H. |
| 318 | ...... do ............... | ♀ | ....do ............... | Oct. 17 | Y. & H. |
| 319 | ...... do ............... | ♂ | ....do ............... | Oct. 17 | Y. & H. |
| 321 | ...... do ............... | ♂ | ....do ............... | Oct. 17 | Y. & H. |
| A7 | ...... do ....:..... (skull ?). | ........ | Alcoholic ........... | ........ | Y. & H. |
| A8 | ...... do ............... | ........ | ....do ............... | ........ | Y. & H. |

| No. | Name. | Sex. | Locality. | Date. | Collector. |
|---|---|---|---|---|---|
| 284 | Catherpes, var. conspersus .. | ♀ ad. | Toquerville, Utah..... | Oct. 13 | H. |
| 304 | ...... do .................... | ♀ ad. | ....do ................ | Oct. 15 | H. |
| 330 | ...... do .................... | ♂ | ....do ................ | Oct. 20 | Y. & H. |
| 83 | Telmatodytes, var. palustris. | ♀ ad. | Provo, Utah .......... | July 25 | Y. & H. |
| 84 | ...... do .................... | ♂ ad. | ....do ................ | July 30 | Y. & H. |
| 208 | ...... do .................... | ♂ | Rush Lake, Utah ..... | Oct. 2 | H. |
| 209 | ...... do .................... | ♀ | ....do ................ | Oct. 2 | H. |
| 306 | ...... do .................... | ♀ | Toquerville, Utah..... | Oct. 15 | H. |
| 259 | Thryotorus bewickii, var. leucogaster. | ♀ ad. | Iron City, Utah....... | Oct. 6 | H. |
| 328 | ...... do .................... | ♂ | Toquerville, Utah..... | Oct. 20 | Y. & H. |
| 356 | ...... do .................... | 0 | Washington, Utah .... | Oct. 23 | H. |
| 369 | Campylorhynchus brunnei-capillus. | ♀ | Saint George, Utah ... | Oct. 27 | Y. & H. |

MOTACILLIDÆ.

*Anthus ludovicianus*, (Gm.)—Tit-Lark.

Quite common in the latter part of summer and in fall in Eastern Nevada and Middle and Southern Utah.

| No. | Name. | Sex. | Locality. | Date. | Collector. |
|---|---|---|---|---|---|
| 211 | Anthus ludovicianus........ | ♀ jun. | Rush Lake, Utah ..... | Oct. 2 | H. |
| 354 | ...... do .................... | ♂ | Washington, Utah.... | Oct. 23 | Y. & H. |
| a9 | ...... do .................... | ....... | Alcoholic....... ...... | .......... | Y. & H. |

SYLVICOLIDÆ.

*Dendroica æstiva*, (Gm.)—Yellow Warbler.

Very abundant at Provo in July, and is the common warbler of all the settlements.

*Dendroica audubonii*, (Towns.)—Audubon's Warbler.

First seen on its way south upon the foot-hills near Gunnison Valley. Afterward met with at different points, usually in small flocks; its habits and call-notes resembling the common eastern Yellow-rump Warbler.

*Geothlypis trichas*, (L.)—Maryland Yellowthroat.

Tolerably abundant at Provo in July; rather rare at other points.

*Geothlypis philadelphia* (Wils.), var. *macgillivrayi*, Aud.—MacGillivray's Warbler.

Secured in mountains of Nevada during the latter part of August, where it appeared tolerably common. Single specimen seen in Damill's Cañon, Utah, in August.

*Icteria virens* (L.), var. *longicauda* Lawr.—Long-tailed Chat.

Common in thickets near Provo; shy and retiring in disposition. Also noticed in Western Utah and Nevada.

*Myiodioctes pusillus,* (Wils.)—Green Black-cap Flycatcher.
Not uncommon in Middle Utah.

*Setophaga ruticilla,* (L.)—Redstart.
Rather common in cottonwood-groves near Provo.

| No. | Name. | Sex. | Locality. | Date. | Collector. |
|---|---|---|---|---|---|
| 55 | Dendroica æstiva | ♀ jun. | Provo, Utah | July 25 | H. & Y. |
| 121 | ...... do | ♀ ad. | ....do | July 25 | H. & Y. |
| 144 | ...... do | ♀ jun. | ....do | July 30 | H. & Y. |
| 145 | ...... do | ♀ ad. | ....do | July 30 | H. & Y. |
| A10 | ...... do | | Alcoholic. | | H. & Y. |
| A11 | ...... do | | ....do | | H. & Y. |
| 152 | Dendroica audubonii | ♂ | Gunnison, Utah | Sept. 8 | H. |
| 216 | ...... do | ♂ | Mormon Spring, Utah. | Oct. 3 | H. |
| 217 | ...... do | ♂ | ....do | Oct. 3 | H. |
| 251 | ...... do | ♂ | Iron City, Utah | Oct. 6 | H. |
| 310 | ...... do | ♂ | Toquerville, Utah | Oct. 16 | H. & Y. |
| 97 | Geothlypis trichas | ♂ ad. | Provo, Utah | July 26 | H. & Y. |
| 177 | ...... do | ♂ ad. | Panquitch, Utah | Sept. 17 | H. |
| A12 | ...... do | | Alcoholic. | | H. & Y. |
| 70 | Geothlypis, var. macgilli- vrayi. | ♀ ad. | Damill's Cañon, Utah. | Aug. 12 | H. |
| A13 | ...... do | | Alcoholic. | | H. & Y. |
| 20 | Icteria, var. longicauda | ♀ ad. | Provo, Utah | July 27 | H. & Y. |
| 146 | ...... do | ♀ jun. | ....do | July 27 | H. & Y. |
| 303 | Myiodioctes pusillus | ♂ ad. | North Creek, Utah | Sept. 26 | H. |
| 122 | Setophaga ruticilla | ♂ ad. | Provo, Utah | July 29 | H. & Y. |

## HIRUNDINIDÆ.

*Hirundo horreorum,* Barton.—Barn-Swallow.

Specimens were secured both in Nevada and in Utah at various points, but it was rather uncommon. Does not differ in any respect from the eastern variety.

*Petrochelidon lunifrons,* (Say.)—Cliff-Swallow.

Observed in Snake Valley, Nevada, and in many localities in Middle and Southern Utah, living in colonies and building their nests at times in inaccessible places in lofty cliffs, and again in places but a few feet above the plain.

*Tachycineta bicolor,* (V.)—White-bellied Swallow.

Found at Fairfield early in August, and at Provo in same month. Rather common.

*Cotyle riparia,* (L.)—Bank-Swallow.

By no means as numerous as the following species, with which it was found associated on the Provo River.

*Stelgidopteryx serripennis,* (And.)—Rough-wing Swallow.

Exceedingly abundant on the Provo River, where they roost in large numbers upon the dead bushes along the banks. So numerous are they and so closely do they sit huddled together, that six individuals were secured at a single shot. They were observed on the wing in pursuit of insects, far into the evening, even when so dark that they could with difficulty be distinguished. Also noticed in Western Utah and Eastern Nevada.

*Progne subis,* (L.)—Purple Martin.

Seen in the vicinity of Salt Lake City and at Camp Douglass. In the middle of July the young were almost able to take wing. Also seen in mountains of Middle Utah.

| No. | Name. | Sex. | Locality. | Date. | Collector. |
|---|---|---|---|---|---|
| 50 | Hirundo horreorum........ | ♂ ad. | Provo, Utah .......... | July 29 | H. |
| 168 | ......do........ ............ | 0 | Fairfield, Utah ....... | Aug. 3 | Y. |
| 90 | Stelgidopteryx serripennis.. | ♂ ad. | Provo, Utah .......... | July 26 | H. & Y. |
| 89 | ......do ........ ....... | ♂ ad. | .... do ............... | July 26 | H. & Y. |
| 95 | ......do .......... ........ | ♂ ad. | .... do ............... | July 26 | H. & Y. |
| 91 | ......do ........ | ♀ jun. | .... do ............... | July 26 | H. & Y. |
| 167 | ......do ........ | 0 | Fairfield, Utah........ | Aug. 3 | Y. |
| A14 | ......do ........ | | Alcoholic ............... | | H. & Y. |
| 96 | Cotyle riparia ............... | 0 | Provo, Utah .......... | July 23 | H. & Y. |

VIREONIDÆ.

*Vireo gilvus,* (V.), var. *swainsoni,* Bd.—Western Warbling Vireo.

Probably not uncommon, though but few were seen. The song of this bird was heard and appeared identical with that of the eastern variety.

*Vireo solitarius,* (Wils.), var. *plumbeus,* Cs.—Western Solitary Vireo.

A single specimen was taken in August near Strawberry Valley. Seen nowhere else in Utah.

| No. | Name. | Sex. | Locality. | Date. | Collector. |
|---|---|---|---|---|---|
| 64 | Vireo, var. swainsoni ........ | ♂ jun. | Wahsatch Mts., Utah . | Aug. 11 | H. |
| B | ...... do ................ | 0 | Meadow Creek, Utah.. | Sept. 15 | Y. |
| 95 | Vireo, var. plumbeus ........ | Jun. | Wahsatch Mts., Utah . | Aug. 16 | H. |

AMPELIDÆ.

*Myiadestes townsendi,* (Aud.)—Townsend's Flycatcher.

Apparently rather rare, not being seen until October 28, when three were noticed in company near Pine Valley. Very shy and retiring, frequenting the hill-sides covered with small cedars, the berries of which constitute the major part of their food in winter when the ground is covered with snow. In some of its habits and motions it closely resembles the bluebirds (*Sialia*). Stationing itself upon the low branches of a tree, it carefully scans the ground, and, perceiving an insect, suddenly darts down and seizing its prey bears it at once to the nearest perch. Have not noticed it catching insects on the wing. It is said to sing always in winter, but this fact was not ascertained.

| No. | Name. | Sex. | Locality. | Date. | Collector. |
|---|---|---|---|---|---|
| 370 | Myiadestes townsendi ...... | ♀ ad. | Pine Valley, Utah .... | Oct. 28 | Y. & H. |
| 402 | ...... do ................ | ♀ jun. | Cove Creek, Utah..... | Nov. 13 | Y. & H. |

## LANIIDÆ.

*Collurio ludovicianus*, (L.), var. *excubitoroides*, Sw.—White-rumped Shrike.

Of frequent occurrence throughout Utah. Subsists largely upon grasshoppers and insects, but occasionally attacks successfully the smaller species of birds and mice. Seen also in Eastern Nevada.

*Collurio borealis*, (V.)—Great Northern Shrike.

Specimen taken and others observed late in the fall in Southern Utah.

| No. | Name. | Sex. | Locality. | Date. | Collector. |
|---|---|---|---|---|---|
| 122 | Collurio, var. excubitoroides. | ♂ jun. | Fairview, Utah | Aug. 22 | H. |
| 115 | ...... do ...... | ♀ jun. | Fountain Green, Utah. | Aug. 20 | H. |
| 375 | ...... do ...... | ♂ ad. | Rush Lake, Utah | Oct. 31 | H. & Y. |
| 410 | ...... do ...... | ♂ ad. | Fillmore, Utah | Nov. 15 | H. |
| 362 | Collurio borealis | ♀ jun. | Saint George, Utah | Oct. 25 | Y. & H. |
| A15 | Collurio excubitoroides | ........ | Alcoholic | ........ | ........ |

### TANAGRIDÆ.

*Pyranga ludoviciana*, (Wils.)—Louisiana Tanager.

A single specimen secured at Provo, where it breeds. Probably not uncommon.

| No. | Name. | Sex. | Locality. | Date. | Collector. |
|---|---|---|---|---|---|
| 126 | Pyranga ludoviciana | ♂ ad. | Provo, Utah | July 29 | H. & Y. |

### FRINGILLIDÆ.

*Carpodacus frontalis*, (Say.)—House-Finch.

Common throughout Utah and Nevada, generally near the settlements. Large flocks seen at Beaver, in the middle of September, searching beneath weeds for seeds.

*Chrysomitris tristis*, (L.)—Yellowbird.

Common. Seen throughout Utah and Eastern Nevada.

*Chrysomitris psaltria*, (Say.)—Arkansas Goldfinch.

Careful search was made for this species in July, but it was not observed until the middle of September in Southern Utah; at this place occasional small flocks were seen frequenting the tall weeds and sunflowers, which latter plant is very common throughout Utah and Nevada; its seeds furnish a nutritious diet for the roving bands of Indians.

*Melospiza melodia*, (Wils.), var. *fallax*, Bd.—Mountain Song-Sparrow.

Common everywhere, and has the same habits, note, and song as the eastern bird.

*Melospiza palustris*, (Wils.)—Swamp-Sparrow.

A single specimen taken at Washington, Utah, in October. This capture affords a valuable fact as far as regards the geographical distribution of this species, as it has never before been taken west of the great plains, its western limit being Eastern Kansas. This being the only specimen taken, it must be regarded as rare, as careful search was made, and hundreds of flocks of sparrows (principally *Zonotrichia*) carefully examined with a view to finding rarities, the fields in the vicinity of Washington being fairly alive with these birds.

*Melospiza lincolni*, (Aud.)—Lincoln's Finch.

Apparently rather rare. Several specimens were taken in a moist meadow in Grass Valley, Utah, September 8, and a pair at Toquerville in the middle of October.

*Spizella pallida*, (Sw.), var. *breweri*, Cass.—Brewer's Sparrow.

Common on the "benches" near Provo, in August. At this time they were in flocks, preparatory to migrating.

*Spizella socialis*, (Wils.), var. *arizonæ*, Cs.

A single specimen taken in Provo Cañon in August, and others in the Wahsatch range of mountains during the migrations in September and October.

*Spizella monticola*, (Gm.)—Tree Sparrow.

A few individuals met with at Beaver about the 1st of November, and was found common at Provo in December.

*Zonotrichia leucophrys*, (Forst.)—White-crowned Sparrow.

A female and young bird were obtained in the Wahsatch Mountains in August, when it breeds abundantly. A single bird was obtained in Southern Utah in October, in a large flock of the following species.

*Zonotrichia leucophrys*, (Forst.), var. *intermedia*, Ridgw.—Western White-crowned Sparrow.

Numerous flocks met with in Southern Utah about the first of October, frequenting the neighborhood of small streams. At this time the preceding species appeared to have departed farther south, as only one specimen was secured; *leucophrys* appearing to be replaced by *intermedia*, which probably winters in the neighborhood of Saint George.

*Poospiza belli*, (Cass.), var. *nevadensis*, Ridgw.—Sage-brush Sparrow.

Numerous specimens obtained of this species, which was first seen near Rush Lake, Utah, October 5. It was observed in small migratory companies of from three to ten, frequenting the sage-brush on desolate plains. Very shy, and was most often seen running with great agility among the bushes; its motions being so quick that it might readily be mistaken for a mouse. In running, its long tail is carried in a perpendicular position, in this respect greatly resembling the wrens. No notes were heard save its single sparrow-like chirp.

*Poœcetes gramineus*, (Gm.), var. *confinus*, Bd.—Grass-Finch.

Common throughout Middle and Southern Utah until October, and, like the preceding species, an inhabitant of the open plains.

*Coturniculus passerinus*, (Wils.), var. *perpallidus*, Ridg.—Western Yellow-winged Sparrow.

A single specimen taken in a cañon near Salina, Gunnison Valley, September 7.

*Passerculus saranna*, (Wils.), var. *alaudinus*, Bp.—Lark-Sparrow.

Common throughout Eastern Nevada and Utah, in the neighborhood of moist places.

*Junco hyemalis*, (L.)—Black Snowbird.

Apparently rare, but a single specimen having been secured at Iron Springs October 4. Never before chronicled from this locality.

*Junco oregonus*, (Towns.)—Oregon Snowbird.

Met with in large flocks in Southern Utah about the middle of October. Also common at Provo in December.

*Passerella townsendi*, (Aud.), var. *schistacea*, Bd.—Slate-colored Sparrow.

A single specimen secured late in July at Provo, Utah. Not seen elsewhere.

*Calamospiza bicolor*, (Towns.)—White-winged Blackbird.

Seen only in Snake Valley, Nevada, and is new to this State.

*Chondestes grammaca*, (Say.)—Lark Finch.

Common throughout Eastern Nevada and Utah; generally found near the water-courses.

*Cyanospiza amœna*, (Say.)—Lazuli Finch.

Very common throughout the Territory of Utah, inhabiting the dense thickets near water-courses. A number of nests were found at Provo in the latter part of July, containing either young or eggs just ready to hatch. (Seen also in Nevada.) These nests were all built upon low thorny bushes, and both nests and eggs resemble those of the Indigo Bird (*C. cyanea*).

*Hedymeles melanocephalus*, (Sw.)—Black-headed Grossbeak.

Probably common throughout Utah, but particularly numerous at Provo, inhabiting the fringes of cottonwood along streams.

*Pipilo maculatus*, Sw., var. *megalonyx* (Bd.)—Spurred Towhee.

Common throughout Nevada and Utah in thickets. Few seen at Provo in December.

*Pipilo abertii*, Bd.—Abert's Towhee.

A pair of these birds, which are not recorded from any locality farther north than Arizona, were secured at Washington and Saint George, Utah. Apparently not uncommon in this locality, as a number of individuals were seen in hedges and scrub. Shy and retiring in disposition, they were difficult to approach.

*Pipilo chlorurus*, (Towns.)—Green-tailed Finch.

Rather common in brush of cañons and mountain-sides throughout Utah.

| No. | Name. | Sex. | Locality. | Date. | Collector. |
|---|---|---|---|---|---|
| 51 | Melospiza melodia, var. fallax | ♀ ad. | Provo, Utah ........... | July 25 | Y. & H. |
| 53 | ...... do ...................... | ♂ ad. | ....do ................. | July 25 | Y. & H. |
| 80 | ...... do ...................... | ♂ ad. | ....do ................. | July 26 | Y. & H. |
| 123 | ...... do ...................... | ♀ ad. | ....do ................. | July 29 | Y. & H. |
| 149 | ...... do ...................... | ♂ ad. | ...do .................. | July 30 | Y. & H. |
| 150 | ...... do ...................... | ♀ ad. | ....do ................. | July 30 | Y. & H. |
| 74 | ...... do ...................... | ♂ jun. | Damill's Cañon, Utah . | Aug. 12 | H. |
| 101 | ...... do ...................... | ♂ jun. | Wahsatch Mts., Utah.. | Aug. 17 | H. |
| 156 | ...... do ...................... | ♀ jun. | Grass Valley, Utah.... | Sept. 10 | H. |
| 157 | ...... do ...................... | ♂ | ...do .................. | Sept. 10 | H. |
| 298 | ...... do ...................... | ♂ | Beaver, Utah ......... | Sept. 25 | H. & Y. |
| 299 | ...... do ...................... | ♂ ad. | ...do .................. | Sept. 25 | H. & Y. |
| 305 | ...... do ...................... | ♂ | North Creek, Utah.... | Sept. 26 | H. & Y. |
| 237 | ...... do ...................... | ♂ | Iron Springs, Utah.... | Oct. 5 | H. |
| 277 | ...... do ...................... | ♂ | ...do .................. | Oct. 10 | H. |
| 303 | ...... do ...................... | ♀ | Toquerville, Utah..... | Oct. 15 | H. |
| 357 | ...... do ...................... | ♀ | Washington, Utah .... | Oct. 24 | H. & Y. |
| 358 | ...... do ...................... | ♀ | ...do .................. | Oct. 23 | H. & Y. |
| 383 | ...... do ...................... | ♂ | Beaver, Utah ......... | Nov. 8 | H. & Y. |
| 412 | ...... do ...................... | ♀ | Fillmore, Utah ........ | Nov. 16 | H. & Y. |
| B5 | ...... do ...................... | ...... | Alcoholic.............. | ......... | H. & Y. |
| 350 | Melospiza palustris.......... | ♂ ad. | Washington, Utah .... | Oct. 23 | H. & Y. |
| 155 | Melospiza lincolni. .......... | ♂ ad. | Grass Valley, Utah.... | Sept. 10 | H. |
| 295 | ...... do ...................... | ♂ ? jun. | ...do .................. | Sept. 14 | H. |
| 9 | Spizella pallida, var. breweri. | ♂ jun. | Provo Cañon, Utah.... | July 31 | H. |
| 10 | ...... do ...................... | ♀ jun. | ...do .................. | July 31 | H. |
| 35 | ...... do ...................... | ♂ ad. | ....do ................. | Aug. 2 | Y. & H. |
| 36 | ...... do ...................... | ♀ ad. | ...do .................. | Aug. 2 | Y. & H. |
| 38 | ...... do ...................... | ♂ jun. | Provo, Utah ........... | Aug. 2 | Y. & H. |
| 145 | ...... do ...................... | ♂ ad. | Salina, Utah .......... | Sept. 5 | H. |
| 178 | ...... do ...................... | ♀ ad. | Panquitch, Utah....... | Sept. 17 | H. |
| 230 | ...... do ...................... | ♂ | Iron Springs, Utah.... | Oct. 4 | H. |
| B2 | ...... do ...................... | ...... | Alcoholic.............. | ......... | H. & Y. |
| B3 | ...... do ...................... | ...... | ...do :................ | ......... | H. & Y. |
| B4 | Spizella socialis, var. arizonae. | ♂ | Provo Cañon, Utah.... | Aug. 11 | H. |
| 153 | ...... do ...................... | ♂ | Wahsatch Mts., Utah.. | Sept. 8 | H. |
| 252 | ...... do ...................... | ♂ ad. | ...do .................. | Oct. 6 | H. |
| 71 | Zonotrichia leucophrys .... | ♂ jun. | Damill's Cañon, Utah . | Aug. 12 | H. |
| 80 | ...... do ................... | ♀ ad. | Strawberry Valley, Utah. | Aug. 13 | H. |
| 214 | ...... do .................... | ♂ jun. | Iron Springs, Utah.... | Oct. 3 | H. |
| 174 | Zonotrichia leucophrys, var. intermedia. | ♂ ad. | Panquitch, Utah ..... | Sept. 17 | H. |
| 172 | ...... do ...................... | ♀ ad. | ...do .................. | Sept. 17 | H. |
| 175 | ...... do ...................... | ♀ jun. | ...do .................. | Sept. 17 | H. |
| 173 | ...... do ...................... | ♀ jun. | ...do .................. | Sept. 17 | H. |
| 211 | ...... do ...................... | ♂ ad. | Iron Springs, Utah.... | Oct. 3 | H. |
| 212 | ...... do ...................... | ♀ ad. | ...do .................. | Oct. 3 | H. |
| 213 | ...... do ...................... | ♂ ad. | ...do .................. | Oct. 3 | H. |
| 215 | ...... do- ..................... | ♀ jun. | ....do ................. | Oct. 3 | H. |
| 223 | ...... do ...................... | ♂ ad. | ...do .................. | Oct. 4 | H. |
| 224 | ...... do ...................... | ♂ ad. | ...do .................. | Oct. 4 | H. |
| 225 | ...... do ...................... | ♂ ad. | ....do ................. | Oct. 4 | H. |
| 227 | ...... do ...................... | ♀ jun. | ....do ................. | Oct. 4 | H. |
| 226 | ...... do ...................... | ♂ ad. | ...do .................. | Oct. 4 | H. |
| 228 | ...... do ...................... | ♀ jun. | ...do .................. | Oct. 4 | H. |
| 229 | ...... do ...................... | ♂ jun. | ...do .................. | Oct. 4 | H. |
| 272 | ...... do ...................... | ♂ jun. | Toquerville, Utah...... | Oct. 4 | H. |
| 253 | ...... do ...................... | ♂ ad. | ...do .................. | Oct. 6 | H. |
| 254 | ...... do ...................... | ♂ ad. | ...do .................. | Oct. 6 | H. |
| 255 | ...... do ...................... | ♀ jun. | ...do .................. | Oct. 6 | H. |
| 270 | ...... do ...................... | ♂ ad. | ...do .................. | Oct. 10 | H. |
| 271 | ...... do ...................... | ♂ ad. | ...do .................. | Oct. 10 | H. |
| 296 | ...... do ...................... | ♀ jun. | ...do .................. | Oct. 14 | H. |
| 316 | ...... do ...................... | ♀ jun. | ...do .................. | Oct. 16 | Y. & H. |
| 317 | ...... do ...................... | ♂ jun. | ...do .................. | Oct. 16 | Y. & H. |

| No. | Name. | Sex. | Locality. | Date. | Collector. |
|---|---|---|---|---|---|
| 297 | Zonotrichia leucophrys, var. intermedia. | ♂ jun. | Toquerville, Utah | Oct. 17 | H. |
| 326 | ...... do | ♂ jun. | ...do | Oct. 19 | H. & Y. |
| 349 | ...... do | ♂ ad. | Washington, Utah | Oct. 23 | H. & Y. |
| 352 | ...... do | ♀ ad. | ....do | Oct. 23 | H. & Y. |
| 356 | ...... do | ♀ jun. | ....do | Oct. 23 | H. & Y. |
| 354 | ...... do | ♂ ad. | ....do | Oct. 24 | H. & Y. |
| 355 | ...... do | ♂ ad. | ....do | Oct. 24 | H. & Y. |
| 116 | ...... do | | Alcoholic | | H. & Y. |
| 117 | .... do | | ....do | | H. & Y. |
| 434 | Spizella monticola | ♀ | Provo, Utah | Nov. 25 | H. |
| 114 | ..... do | | Alcoholic | | H. & Y. |
| 221 | Poospiza bellii, var. nevadensis. | ♀ | Iron Springs, Utah | Oct. 4 | H. |
| 220 | ...... do | ♀ | ....do | Oct. 4 | H. |
| 222 | ...... do | ♂ | ....do | Oct. 4 | H. |
| 233 | ...... do | ♂ | ....do | Oct. 4 | H. |
| 234 | ......, do | ♂ | ....do | Oct. 4 | H. |
| 236 | ...... do | ♂ (?) | ....do | Oct. 5 | H. |
| 235 | ...... do | ♀ | ....do | Oct. 5 | H. |
| 314 | ...... do | ♂ | Toquerville, Utah | Oct. 16 | H. |
| 315 | ...... do | ♂ | ....do | Oct. 16 | H. |
| 325 | ...... do | ♂ | ....do | Oct. 19 | H. |
| 371 | ...... do | ♂ | Saint George, Utah | Oct. 28 | Y. & H. |
| 372 | ...... do | ♂ | ....do | Oct. 28 | Y. & H. |
| 81 | Poœcetes gramineus, var. confinis. | ♀ ad. | Strawberry Valley, Utah. | Aug. 13 | H. |
| 100 | ...... do | ♀ jun. | Panquitch, Utah | Aug. 17 | H. |
| 144 | ...... do | ♂ jun. | Salina, Utah | Sept. 5 | H. |
| 152 | ...... do | ♂ jun. | Grass Valley, Utah | Sept. 10 | H. |
| 292 | ...... do | ♂ ad. | Beaver, Utah | Sept. 24 | Y. & H. |
| 294 | ...... do | ♀ ad. | Toquerville, Utah | Oct. 14 | H. |
| 118 | ...... do | | Alcoholic | | H. & Y. |
| 149 | Coturniculus passerinus, var. perpallidus. | ♀ jun. | Gunnison, Utah | Sept. 7 | H. |
| 123 | Passerculus savanna, var. alaudinus. | ♀ jun. | Thistle Valley, Utah | July 20 | H. |
| 78 | ...... do | ♂ jun. | Provo, Utah | July 26 | Y. & H. |
| 79 | ...... do | ♂ ad. | ....do | July 26 | Y. & H. |
| 81 | ...... do | ♂ ad. | ....do | July 26 | Y. & H. |
| 39 | ...... do | ♂ ad. | ....do | Aug. 2 | Y. & H. |
| 39a | ...... do | ♀ ad. | ....do | Aug. 2 | H. |
| 40 | ...... do | ♂ ad. | ....do | Aug. 2 | H. & Y. |
| 232 | Junco hyemalis | ♀ jun. | Iron Springs, Utah | Oct. 4 | H. |
| 324 | Junco oregonus | ♂ | Toquerville, Utah | Oct. 19 | H. & Y. |
| 323a | ...... do | ♂ | ....do | Oct. 19 | H. & Y. |
| 353 | ...... do | ♀ | Washington, Utah | Oct. 23 | H. & Y. |
| 435 | ...... do | ♂ | Provo, Utah | Nov. 25 | H. & Y. |
| 460 | ...... do | ♂ | ....do | Nov. 29 | H. & Y. |
| 136 | Passerella townsendi, var. schistacea. | ♀ ad. | ....do | July 20 | H. & Y. |
| 21 | Carpodacus frontalis | ♂ ad. | ....do | Aug. 1 | H. & Y. |
| 52 | ...... do | ♀ jun. | ....do | July 25 | H. & Y. |
| 124 | ...... do | ♀ ad. | ....do | July 29 | H. & Y. |
| 41 | ...... do | ♀ ad. | ....do | Aug. 2 | H. & Y. |
| 42 | ...... do | ♂ ad. | ....do | Aug. 2 | H. & Y. |
| 176 | ...... do | 0 | Panquitch, Utah | Sept. 17 | H. |
| 350 | ...... do | ♂ jun. | Washington, Utah | Oct. 24 | H. & Y. |
| 296 | ...... do | ♀ jun. | Beaver, Utah | Sept. 25 | H. & Y. |
| 297 | ...... do | ♂ | ....do | Sept. 25 | H. & Y. |
| 309 | ...... do | ♀ jun. | North Creek, Utah | Sept. 26 | H. & Y. |
| 231 | ...... do | ♀ jun. | Iron Springs, Utah | Oct. 4 | H. |
| 256 | ...... do | ♂ ad. | Iron City, Utah | Oct. 6 | H. |
| 257 | ...... do | ♀ jun. | ....do | Oct. 6 | H. |
| 346 | ...... do | ♂ ad. | Washington, Utah | Oct. 23 | H. & Y. |
| 347 | ...... do | ♀ | ....do | Oct. 23 | H. & Y. |

2 O S

| No. | Name. | Sex. | Locality. | Date. | Collector. |
|---|---|---|---|---|---|
| 348 | Carpodacus frontalis......... | ♀ jun. | Washington, Utah.... | Oct. 23 | H. & Y. |
| a20 | ..... do ...................... | ...... | Alcoholic............. | | H. & Y. |
| 127 | Chrysomitris tristris......... | ♂ | Provo, Utah .......... | July 29 | H. & Y. |
| 143 | ..... do ..................... | ♂ ad. | ....do ............... | July 30 | H. & Y. |
| 478 | ..... do ..................... | ♂ jun. | ....do ............... | Dec. 1 | H. & Y. |
| 273 | Chrysomitris psaltria........ | ♂ ad. | Washington, Utah .... | Oct. 10 | H. & Y. |
| 274 | ..... do ..................... | ♀ jun. | ....do ............... | Oct. 10 | H. & Y. |
| 275 | ..... do ..................... | ♂ jun. | ....do ............... | Oct. 10 | H. & Y. |
| 356 | ..... do ..................... | ♂ jun. | ....do ............... | Oct. 23 | H. & Y. |
| 365 | ..... do ..................... | ♀ jun. | Saint George, Utah.... | Oct. 26 | H. & Y. |
| 366 | ..... do ..................... | ♂ jun. | ....do ............... | Oct. 26 | H. & Y. |
| 367 | ..... do ..................... | ♂ jun. | ....do ............... | Oct. 26 | H. & Y. |
| 368 | ..... do ..................... | ♂ jun. | ....do ............... | Oct. 26 | H. & Y. |
| 118 | Chondestes grammaca...... | ♀ jun. | Provo, Utah .......... | July 24 | H. & Y. |
| 119 | ..... do ..................... | ♀ jun. | ....do ............... | July 31 | H. & Y. |
| 180 | ..... do ..................... | ♀ jun. | ....do ............... | July 31 | H. & Y. |
| 33 | ..... do ..................... | ♂ jun. | ....do ............... | Aug. 2 | H. & Y. |
| 34 | ..... do ..................... | ♀ ad. | ....do ............... | Aug. 3 | H. & Y. |
| 47 | ..... do ..................... | ♀ jun. | ....do ............... | Aug. 3 | H. & Y. |
| 108 | Cyanospiza amœna.......... | ♂ ad. | ....do ............... | July 29 | H. & Y. |
| 125 | ..... do ..................... | ♀ ad. | ....do ............... | July 29 | H. & Y. |
| 148 | ..... do ..................... | ♀ ad. | ....do ............... | July 30 | H. & Y. |
| 22 | ..... do ..................... | ♂ ad. | ....do ............... | Aug. 1 | H. & Y. |
| 23 | ..... do ..................... | ♀ ad. | ....do ............... | Aug. 1 | H. & Y. |
| 102 | ..... do ..................... | ♀ ad. | ....do ............... | Aug. 17 | H. & Y. |
| B1 | ..... do ..................... | ...... | Alcoholic............. | | H. & Y. |
| 25 | Hedymeles melanocephalus . | ♂ jun. | Provo, Utah .......... | July 24 | H. & Y. |
| 131 | ..... do ..................... | ♂ ad. | ....do ............... | July 24 | H. & Y. |
| 26 | ..... do ..................... | ♂ ad. | ....do ............... | July 29 | H. & Y. |
| 114 | ..... do ..................... | ♀ jun. | ....do ............... | July 29 | H. & Y. |
| 134 | ..... do ..................... | ♀ jun. | ....do ............... | July 30 | H. & Y. |
| 115 | ..... do ..................... | ♀ ad. | ....do ............... | July 29 | H. & Y. |
| 116 | ..... do ..................... | ♀ jun. | ....do ............... | July 29 | H. & Y. |
| 117 | ..... do ..................... | ♂ jun. | ....do ............... | July 29 | H. & Y. |
| 133 | ..... do ..................... | ♀ jun. | ....do ............... | July 29 | H. & Y. |
| 135 | ..... do ..................... | ♀ ad. | ....do ............... | July 30 | H. & Y. |
| 132 | ..... do , ................... | ♀ jun. | ....do ............... | July 30 | H. & Y. |
| 112 | Pipilo maculatus, var. me-galonyx. | ♂ ad. | ....do ............... | July 29 | H. & Y. |
| 110 | ..... do ..................... | ♀ ad. | ....do ............... | July 29 | H. & Y. |
| 113 | ..... do ..................... | ♀ ad. | ....do ............... | July 29 | H. & Y. |
| 118 | ..... do ..................... | ♀ jun. | ....do ............... | July 30 | H. & Y. |
| 139 | ..... do ..................... | ♂ ad. | ....do ............... | July 30 | H. & Y. |
| 140 | ..... do ..................... | ♀ jun. | ....do ............... | July 30 | H. & Y. |
| 111 | ..... do ..................... | ♀ ad. | ....do ............... | Aug. 1 | H. & Y. |
| 306 | ..... do ..................... | ♂ ad. | North Creek, Utah .... | Sept. 26 | H. & Y. |
| 306a | ..... do ..................... | ♂ ad. | ....do ............... | Sept. 26 | H. & Y. |
| 329 | ..... do ..................... | ♂ ad. | Toquerville, Utah..... | Oct. 24 | H. & Y. |
| 459 | ..... do ..................... | ♀ | Provo, Utah .......... | Nov. 30 | H. & Y. |
| B9 | ..... do ..................... | ...... | Alcoholic............. | | H. & Y. |
| 345 | Pipilo aberti................. | ♂ jun. | Washington, Utah .... | Oct. 22 | Y. & H. |
| 364 | ..... do ..................... | ♀ jun. | Saint George, Utah ... | Oct. 22 | Y. & H. |
| 99 | Pipilo chlorurus............. | ♂ ad. | Wahsatch Mts., Utah.. | Aug. 17 | H. |
| 148 | ..... do ..................... | ♀ ad. | Gunnison, Utah....... | Sept. 7 | H. |
| C | ..... do ..................... | Jun. | Meadow Creek, Utah.. | Sept. 15 | Y. |

## ALANDIDÆ.

*Eremophila alpestris* (Forst).—Horned Lark.

The young in nesting-plumage obtained in Western Utah August 1, and were very abundant in the sage-brush plains. After the 1st of September frequently met with in small scattered flocks.

| No. | Name. | Sex. | Locality. | Date. | Collector. |
|---|---|---|---|---|---|
| 158 | Eremophila alpestris ....... | ♂ jun. | Fairfield, Utah........ | Aug. 1 | Y. |
| 159 | ...... do ................... | ♂ jun. | ....do ................ | Aug. 1 | Y. |
| 160 | ...... do ................... | ♂ jun. | ....do ................ | Aug. 1 | Y. |
| 142 | ...... do ................... | ♂ jun. | Gunnison, Utah....... | Sept. 5 | H. |
| 143 | ...... do ................... | ♂ jun. | ....do ................ | Sept. 5 | H. |
| 351 | ...... do ................... | ♂ | Beaver, Utah ......... | Nov. 7 | Y. & H. |
| B11 | ...... do ................... | ...... | Alcoholic............. | ......... | Y. & H. |

## ICTERIDÆ.

*Dolichonyx oryzivorus*, (L.)—Bobolink.

Rather common in fields in the vicinity of Provo, Utah. The parent birds were noticed feeding their young, scarcely fledged, July 25.

*Molothrus pecoris*, (Gm.)—Cowbird.

Two specimens secured in Provo Cañon in July.

*Agelaius phœniceus*, (L.)—Red-winged Blackbird.

Exceedingly numerous in the marshes throughout Utah and Nevada. Immense flocks were noticed near Provo in December. This and the two following species are cordially detested by the farmers, owing to the great damage done to the crops.

*Xanthocephalus icterocephalus*, (Bonap.)—Yellow-headed Blackbird.

Large flocks seen at Provo in July and in Eastern Nevada in August. A few seen November 15 in Middle Utah, associated with large flocks of Red-winged Blackbirds.

*Scolecophagus cyanocephalus*, (Wagl.)—Brewer's Blackbird.

Most numerous of the blackbirds; large flocks met with throughout Utah and Nevada, frequenting alike fields and marshes. An albino of this species, the upper parts being nearly all white, was seen at Beaver.

*Sturnella magna*, L., var. *neglecta* Aud.—Western Lark.

Abundant in fields near settlements throughout Utah and Eastern Nevada. Probably resident in Southern Utah.

*Icterus bullocki*, (Sw.)—Bullock's Oriole.

A single individual seen at Provo, and one shortly afterward in middle of August in the Wahsatch Mountains. Probably migrates early, as nests presumably of this species were found, but no birds seen after this time.

| No. | Name. | Sex. | Locality. | Date. | Collector. |
|---|---|---|---|---|---|
| 72 | Dolichonyx oryzivorus...... | ♂ ad. | Provo, Utah ......:... | July 21 | H. & Y. |
| 7 | Molothrus pecoris.......... | ♀ juu. | Provo Cañou, Utah ... | July 31 | H. |
| 19 | ...... do .................. | ♂ jnn. | ....do ................. | Aug. 1 | H. |
| 406 | Agelaius phœniceus ........ | ♂ ad. | Cove Creek, Utah...... | Nov. 15 | H. & Y. |
| 407 | ...... do .................. | ♂ ad. | ....do ................. | Nov. 15 | H. & Y. |
| 408 | ...... do .................. | ♂ ad. | ....do ................. | Nov. 15 | H. & Y. |
| 461 | ...... do .................. | ♂ ad. | Provo, Utah .......... | Nov. 30 | H. & Y. |
| 409 | Xanthocephalus icteroce- | ♂ ad. | Cove Creek, Utah..... | Nov. 15 | H. & Y. |
| | phalus. | | | | |
| 30 | Scolecophagus cyanoce- | ♀ jnn. | Provo, Utah .......... | Aug. 2 | H. |
| | phalus. | | | | |
| 32 | ...... do .................. | ♂ jnn. | ....do ................. | Aug. 3 | H. |
| 31 | ...... do .................. | ♀ jnn. | ....do ................. | Aug. 2 | H. |
| 46 | ...... do .................. | ♂ jnn. | ....do ................. | Aug. 2 | H. |
| 165 | ...... do .................. | ♂ ad. | Harmony, Utah....... | Sept. 15 | H. |
| 166 | ...... do .................. | ♂ jnn. | ....do ................. | Sept. 15 | H. |
| 276 | ...... do .................. | ♂ ad. | Iron City, Utah....... | Oct. 10 | H. |
| 291 | ...... do .................. | ♂ ad. | Harmony, Utah....... | Sept. 15 | H. |
| 295 | ...... do .................. | ♀ ad. | Beaver, Utah ........ | Sept. 25 | Y. & H. |
| 363 | ...... do .................. | ♂ ad. | Cove Creek, Utah..... | Oct. 26 | Y. & H. |
| 181 | Sturnella magna, var. ne- | ♀ jnn. | Panquitch, Utah ..... | Sept. 12 | H. |
| | glecta. | | | | |
| 344 | ...... do .................. | ♂ jnn. | Washington, Utah .... | Oct. 22 | H. & Y. |
| 351a | ...... do .................. | ♂ jnn. | ....do ................. | Oct. 23 | H. & Y. |
| 352 | ...... do .................. | ♀ jnn. | ....do ................. | Oct. 23 | H. & Y. |

CORVIDÆ.

*Corvus corax*, L., var. *carnivorus*, Bartr.—American Raven.

The most common and characteristic bird of Nevada and Utah; particularly numerous in the vicinity of cattle-ranges. It is variable in its disposition, at times very shy and at others permitting the closest approach. Not generally found above the plains and foot-hills.

*Corvus americanus*, Aud.—Common Crow.

Comparatively rare; met with only at Provo, where a number were seen at different times. Said by the settlers to have appeared within a few years.

*Picicorvus columbianus*, (Wils.)—Clark's Crow.

Not observed until September 8, when a pair were noticed at Otter Creek, Middle Utah. From this time until the middle of October it was seen almost daily, singly and in flocks. It was invariably on the wing, flying from side to side of the mountains, generally to the numerous cedars, and uttering its peculiar notes, which consist of a succession of short rattling cries. Owing to its singular uneasiness, almost akin to shyness, none were secured. Its flight is undulating and its habits much resemble those of the woodpeckers (*Picidæ*).

*Pica melanoleuca*, V., var. *hudsonica* Sab.—Magpie.

Numerous in mountains and plains of Eastern Nevada and Utah. Shy and difficult of approach. Its voice is singularly flexible, and capable of producing a variety of sounds, from the guttural chuckle to the softest whistle. Resident throughout the year.

*Cyanurus stelleri*, (Gm.), var. *macrolopha* Bd.—Long-crested Jay.

Apparently confined to mountains and cañons. Specimens taken at Provo in July and November. Winters.

*Cyanocitta floridana*, (Bartr.), var. *woodhousii*, Bd.—Woodhouse's Jay.

Common and somewhat abundant in Nevada and Utah; generally found in bushes along streams. In habits and notes, little or no difference could be detected from those of the Florida Jay (*C. floridana*).

*Gymnokitta cyanocephala* Pr. Max.—Maximillian's Jay.

Common in mountains and foot-hills of Nevada and Utah in the vicinity of cedars, the gum of which in all the specimens taken was found adhering to the feathers. At the season when taken, September until December, strictly gregarious.

| No. | Name. | Sex. | Locality. | Date. | Collector. |
|---|---|---|---|---|---|
| 3-2 | Corvus carnivorus | ♀ ad. | Beaver, Utah | Nov. 8 | |
| 112 | Pica melanoleuca, var. hudsonica. | ♂ jun. | Fountain Green, Utah. | Aug. 20 | H. |
| 302 | ...... do | ♀ jun. | North Creek, Utah | Sept. 26 | H. & Y. |
| 411 | ...... do | ♀ jun. | Fillmore, Utah | Nov. 16 | H. & Y. |
| 5 | Cyanurus stelleri, var. macrolopha. | ♀ jun. | Provo Cañon, Utah | July 31 | H. |
| 6 | ...... do | ♂ jun. | ....do | July 31 | H. |
| 462 | ...... do | ♂ ad. | ....do | Nov. 30 | H. & Y. |
| 463 | ...... do | ♀ ad. | ....do | Nov. 30 | H. & Y. |
| 147 | Cyanocitta floridana, var. woodhousii. | ♂ | Gunnison, Utah | Sept. 7 | H. |
| 248 | ...... do | ♂ | Iron City, Utah | Oct. 5 | H. |
| 293 | ...... do | | Beaver, Utah | Sept. 25 | Y. & H. |
| 260 | ...... do | ♂ | Fillmore, Utah | Sept. 4 | Y. |
| 398 | ...... do | ♂ | Pine Creek, Utah | Nov. 12 | Y. & H. |
| 399 | ...... do | ♂ | ....do | Nov. 12 | Y. & H. |
| 400 | ...... do | ♀ | ....do | Nov. 12 | Y. & H. |
| 401 | ...... do | ♂ | ....do | Nov. 12 | Y. & H. |
| 464 | ...... do | ♂ | Provo, Utah | Nov. 25 | Y. & H. |
| 250 | ...... do | ♂ | Iron City, Utah | Oct. 6 | H. |
| 285 | Gymnokitta cyanocephala.. | ♀ ad. | Beaver, Utah | Sept. 24 | Y. & H. |
| 86 | ...... do | ♀ ad. | ....do | Sept. 24 | Y. & H. |
| 87 | ...... do | ♂ ad. | ....do | Sept. 24 | Y. & H. |
| 88 | ...... do | ♀ ad. | ....do | Sept. 24 | Y. & H. |
| 376 | ...... do | ♂ ad. | ....do | Oct. 31 | Y. & H. |
| 77 | ...... do | ♂ ad. | ....do | Oct. 31 | Y. & H. |
| 78 | ...... do | ♀ ad. | ....do | Oct. 31 | Y. & H. |
| 79 | ...... do | ♂ ad. | ....do | Nov. 3 | Y. & H. |

TYRANNIDÆ.

*Tyrannus carolinensis*, (L.)—King-bird.

Not very common, except in the vicinity of Provo River, where it was taken.

*Tyrannus verticalis*, Say.—Arkansas Flycatcher.

Quite common at Provo. A nest found on end of cottonwood limb projecting over the water was composed of cottonwood-down and grasses, lined with a few hairs, and presenting a bulky appearance. The young birds, though able to fly, were being fed by their parents at this date, July 26.

*Empidonax pusillus* (Sw.)—Little Flycatcher.

Exceedingly numerous near Provo River in willow-thickets, sparingly so in Eastern Nevada. Very quick and nervous in its movements, constantly crossing and recrossing the river and catching insects. The single "whit," which is ever repeated, is strongly suggestive of the note of the Least Flycatcher (*E. minimus*), while the song may be compared to that of the eastern Phœbe (*Sayornis fuscus*). A nest found July 27 in a small willow, 3 feet from the ground, was a rather loose structure, composed of grasses, with a lining of a few hairs. This contained newly-hatched young. Eggs white, sprinkled with reddish-brown. A comparison of the large series taken shows considerable variation in size, especially as regards the bills.

*Empidonax hammondii*, Bd.—Hammond's Flycatcher.

Two specimens only secured, one from near Beaver River, and the other procured by Lieutenant Hoxie twenty-five miles from Fillmore, Utah. This gentleman stated that he saw numbers of these birds in a cave in company with swallows.

*Empidonax obscurus*, (Sw.)—Wright's Flycatcher.

Two specimens secured, one in Eastern Nevada, the other on a mountain-side near Provo covered with scrub. Not common.

*Contopus borealis*, (Sw.)—Olive-sided Flycatcher.

Seen upon several occasions in the heavy pine-timber of the Wahsatch. Seems not to differ in coloration from eastern specimens.

*Contopus virens*, (L.), var. *richardsonii* Sw.—Short-legged Pewee.

Seen in same localities as preceding. Apparently rather common.

| No. | Name. | Sex. | Locality. | Date. | Collector. |
|---|---|---|---|---|---|
| 48 | Tyrannus carolinensis | ♀ ad. | Provo, Utah | July 25 | Y. & H. |
| 57 | Tyrannus verticalis | ♂ ad. | ....do | July 26 | Y. & H. |
| 70 | ...... do | ♀ ad. | ....do | July 26 | Y. & H. |
| 103 | ...... do | ♂ ad. | ....do | July 27 | Y. & H. |
| 104 | ...... do | | ....do | July 27 | Y. & H. |
| 35 | Empidonax pusillus | ♀ ad. | ....do | July 25 | Y. & H. |
| 36 | ...... do | ♀ ad. | ....do | July 25 | Y. & H. |
| 37 | ...... do | ♂ ad. | ....do | July 25 | Y. & H. |
| 38 | ...... do | ♂ ad. | ....do | July 25 | Y. & H. |
| 39 | ...... do | ♂ ad. | ....do | July 25 | Y. & H. |
| 40 | ...... do | ♂ ad. | ....do | July 25 | Y. & H. |
| 41 | ...... do | ♂ ad. | ....do | July 25 | Y. & H. |
| 42 | ...... do | ♀ | ....do | July 25 | Y. & H. |
| 43 | ...... do | ♀ ad. | ....do | Aug. 2 | H. |
| 44 | ...... do | ♀ ad. | ....do | July 25 | H. & Y. |
| 43a | ...... do | ♂ ad. | ....do | July 25 | H. & Y. |
| 48 | ...... do | ♂ ad. | ....do | July 29 | H. & Y. |
| 49 | ...... do | ♂ ad. | ....do | Aug. 3 | H. |
| 27 | ...... do | ♂ ad. | ....do | Aug. 1 | H. |
| 73 | ...... do | ♀ ad. | ....do | July 26 | H. & Y. |
| 74 | ...... do | ♂ ad. | ....do | July 26 | H. & Y. |
| 75 | ...... do | ♂ ad. | ....do | July 26 | H. & Y. |
| 75a | ...... do | ♂ ad. | ....do | July 26 | H. & Y. |
| 28 | ...... do | ♀ ad. | ....do | Aug. 3 | H. |
| 85 | ...... do | ♀ ad. | ....do | July 29 | H. & Y. |
| 120 | ...... do | ♀ ad. | ....do | July 29 | H. & Y. |
| 98 | ...... do | ♂ ad. | ....do | July 27 | H. & Y. |
| 151 | ...... do | ♂ ad. | ....do | July 30 | H. & Y. |
| 96 | ...... do | ♀ juv. | Wahsatch, Utah | Aug. 16 | H. |
| 270 | Empidonax hammondii | ♂ | Beaver, Utah | Sept. 22 | Y. & H. |
| D | ...... do | | Cedar, Utah | Oct. — | Hoxie. |
| 62 | Empidonax obscurus | | Provo, Utah | Aug. 9 | H. |
| E | ...... do | | Snake Creek, Nev. | Aug. 9 | Y. |
| 73 | Contopus borealis | ♀ ad. | Daniell's Cañon, Utah. | Aug. 12 | H. |
| 72 | Contopus virens, var. richardsonii. | ♂ ad. | ....do | Aug. 12 | H. |
| 133 | ...... do | ♂ ad. | Wahsatch, Utah | Aug. 25 | H. |

## ALCEDINIDÆ.

*Ceryle alcyon*, (L.)—Belted Kingfisher.
Common on streams throughout Utah.

| No. | Name. | Sex. | Locality. | Date. | Collector. |
|-----|-------|------|-----------|-------|------------|
| 87 | Ceryle alcyon ............... | ♂ ad. | Provo, Utah .......... | July 26 | Y. & H. |

## CAPRIMULGIDÆ.

*Antrostomus nuttallii*, (Aud.)—Poor-will.

Several individuals, believed to be of this species, were seen at various times in Southern Utah, having the peculiar flight in the day-time which is characteristic of this bird.

*Chordeiles popetue*, (V.), var. *henryi* Cass.—Western Night-Hawk.

Very common, especially in Western Utah and Eastern Nevada. Unlike our eastern species, which generally commences to hunt insects at early dusk, this bird was almost invariably seen to commence its repast between three and four in the afternoon.

## CYPSELIDÆ.

*Panyptila saxatilus* (Woodh.)—White-throated Swift.

Noticed on one occasion only while passing the divide between Gunnison and Grass Valley. Several individuals were noticed high in air, but keeping well out of range.

## TROCHILIDÆ.

*Trochilus alexandri*, Bourc and Muls.—Black-chinned Humming-bird.

Observed in cañons in Nevada and at Provo, where it was very numerous, it being the only species of humming-bird taken. Quite common throughout the Territory. A nest found at Provo, July 29, contained two eggs nearly hatched; it was placed in a notch of a cottonwood branch, 12 feet from the tree, and formed of cotton from this tree.*

| No. | Name. | Sex. | Locality. | Date. | Collector. |
|-----|-------|------|-----------|-------|------------|
| 129 | Trochilus alexandri......... | ♀ ad. | Provo, Utah .......... | July 29 | Y. & H. |
| 128 | ...... do ...... ...... | ♂ ad. | .... do ................. | July 29 | Y. & H. |
| 130 | ...... do ...... ...... | ♀ ad. | .... do ................. | July 29 | Y. & H. |
| 131 | ...... do ...... ...... | ♂ ad. | .... do ................. | July 30 | Y. & H. |
| 132 | ...... do ...... ...... | ♀ jun. | .... do ................. | July 30 | Y. & H. |

* It may be mentioned in this connection that while traveling over an Indian trail leading from Long Valley to Shonesburgh, Southern Utah, a pair of humming-birds were noticed at the bottom of a cañon some distance below the trail. At this distance they appeared on the back of a greenish-yellow color, with black stripes in the side of the individual supposed to be the male. An effort was made to secure them but failed. This note is given for what it is worth.—H. C. YARROW.

CUCULIDÆ.

*Geococcyx californianus*, (Less.)

Evidence was obtained from the settlers of its occurrence at Saint George, Southern Utah.

PICIDÆ.

*Colaptes mexicanus*, Sw.—Red-shafted Flicker.

Owing to the general absence of timber, none of the *Picidæ* were common except at a few localities among the mountains. This species is very generally distributed throughout the Territories of Utah and Nevada. At Provo, in July, but few individuals were seen, but in December at this place they were very common. Nests often seen in holes in banks of streams.

*Sphyropicus varius*, (L.), var. *nuchalis*, Bd.—Red-throated Woodpecker.

A pair taken at Toquerville in October, and a few individuals were seen in aspen-groves in the Wahsatch Mountains.

*Melanerpes torquatus* (Wils.)—Lewis's Woodpecker.

A single specimen taken in Beaver Cañon in September, but no doubt this and the preceding species are common in the heavy timber of the mountains.

*Picus villosus*, (L.), var. *harrisii* Aud.—Harris's Woodpecker.

A single male bird secured in Grass Valley in September.

*Picus pubescens*, (L.), var. *gairdneri* Aud.—Gairdner's Woodpecker.

An individual believed to be of this species was seen at Provo, November 27.

| No. | Name. | Sex. | Locality. | Date. | Collector. |
|---|---|---|---|---|---|
| 45 | Colaptes mexicanus ........ | ♂ jun. | Provo, Utah .......... | Aug.  3 | H. |
| 360 | ...... do ..................... | ♀ ad. | Washington, Utah .... | Oct.  24 | H. & Y. |
| 468 | ....... do ................... | ♂ ad. | Provo, Utah .......... | Nov. 30 | H. & Y. |
| 469 | ...... do ..................... | ♂ ad. | .... do ................. | Nov. 30 | H. & Y. |
| 486 | ...... do ..................... | ♂ ad. | .... do ................. | Dec.  1 | H. & Y. |
| 283 | Sphyropicus varius, var. nuchalis | ♂ jun. | Toquerville, Utah..... | Oct.  13 | H. |
| 307 | ...... do ..................... | ♂ jun. | Virgin City, Utah..... | Oct.  14 | Y. |
| 269 | Melanerpes torquatus....... | ♂ jun. | Beaver, Utah ......... | Sept. 22 | Y. & H. |
| 160 | Picus villosus, var. harrisii . | ♂ | Grass Valley, Utah.... | Sept. 10 | H. |

STRIGIDÆ.

*Otus vulgaris*, (L.), var. *wilsonianus* (Less.)—Long-eared Owl.

Probably the most common of this family. A colony of perhaps a dozen individuals met with in a cedar-grove in Grass Valley, Eastern Utah, in September. A number of specimens here obtained.

A favorite abode of this species appears to be the heavy brush found in all the streams issuing from the mountains, which in many cases, indeed, from a lack of heavier timber, constitutes their sole resort.

On a stream near Fillmore, six of these owls were taken within a radius of half a mile, and many of their old nests being seen, there seems no reason to doubt but that this was a favorite breeding-ground.

*Speotyto cunicularia*, var. *hypugœa* Bp. (Mol.)—Prairie-Owl.

Seen in but two localities in Utah, Dog Valley and near Pauquitch Lake. Not very numerous, living with prairie-dogs.

| No. | Name. | Sex. | Locality. | Date. | Collector. |
|---|---|---|---|---|---|
| 154 | Otus vulgaris, var. wilsonianus | ♀ ad. | Grass Valley, Utah.... | Sept. 10 | H. |
| 159 | ...... do ................... | ♂ ad. | ....do ........... | Sept. 10 | H. |
| 421 | ...... do ................... | ♂ ad. | Fillmore............ | Nov. 18 | H. & Y. |
| 422 | ...... do .. ............... | ♂ ad. | ....do ............ | Nov. 18 | H. & Y. |
| 423 | ...... do ............... | ♂ ad. | ....do ....... | Nov. 18 | H. & Y. |
| 426 | ...... do .. ............... | ♀ ad. | ....do ....... | Nov. 18 | H. & Y. |

FALCONIDÆ.

*Falco sparverius*, L.—Sparrow-Hawk.

Very common in Nevada and Utah. Seen frequently in the mountains; subsists largely upon grasshoppers.

*Nisus fuscus*, (Gm.)—Sharp-shinned Hawk.

Not uncommon in Nevada and Utah. A beautiful adult pair were taken in Beaver Cañon September 24. Upon one occasion, while watching a pair of doves feeding upon the ground, a female of this species made a daring and successful swoop upon one of them, passing within a few feet of the observer's head. As a further illustration of the bravery and hardihood with which this bird pursues its prey, it may be mentioned that one was observed in the town of Pauq-uitch eagerly pursuing a common pigeon, apparently oblivious of the presence of spectators, who, for some time, vainly endeavored to drive it away. Such was its determination that it actually followed the pigeon into a deserted house, but was finally obliged to retire without accomplishing its object.

*Buteo borealis*, (Gm), var. *calurus* Cass.

A fine adult female was secured at Otter Creek, Utah, in September. Observed at intervals during the entire season, generally in the mountains.*

*Archibuteo lagopus*, (Brum.), var. *sancti-johannis* Gm.—Black Hawk.

Although seen several times in the mountains during the summer, none were obtained until at Provo, where it was the most numerous of the hawks. At this place, from November 25 until December 4, no less than eleven specimens were taken, representing the bird in all stages of plumage. On foot it was extremely difficult to approach this hawk, but it could be ridden up to with ease; most of the specimens being shot in this way from the back of a mule.

Utah Lake and the surrounding marshes attract multitudes of water-

---

* It may be mentioned that in the road from Saint George to Beaver, late in the season, several *Buteos* were observed not *calurus*, but probably allied to *lineatus*, as it resembled it in form and flight.

fowl, and this undoubtedly explains in part the abundance of hawks at this season, since wounded and disabled ducks must form no inconsiderable part of their food. Its manner of hunting much resembles the following species, and like it subsists to a certain extent upon mice, which are very numerous in the rushes. In the stomachs of every individual captured was found the remains of these little animals.

*Circus cyaneus*, (L.), var. *hudsonius* L.—Marsh-Hawk.

Frequently observed in the lowlands during the trip nearly as abundant as the preceding at Provo in the fall. To be seen at all hours of the day, sweeping over the tops of the marsh-rushes in search of mice. In several instances these birds were decoyed within gunshot by the collector hiding in the rushes and imitating the squeak of a mouse.

*Haliaëtus leucocephalus*, (L.)—Bald Eagle.
Of frequent occurrence in the lowlands.

*Aquila chrysaëtos*, (L.), var. *canadensis* L.—Golden Eagle.
Of frequent occurrence in the mountains.

| No. | Name. | Sex. | Locality. | Date. | Collector. |
|---|---|---|---|---|---|
| 97 | Falco sparverius ........... | ♀ ad. | Wahsatch Mts., Utah.. | Aug. 16 | H. |
| 268 | ...... do ..................... | ♂ ad. | Beaver, Utah ......... | Sept. 22 | Y. & H. |
| 290 | ...... do ..................... | ♂ ad. | ....do ................. | Sept. 24 | Y. & H. |
| 184 | Buteo borealis, var. calurus . | ♀ ad. | Otter Creek, Utah..... | Sept. 19 | H. |
| 284 | Nisus fuscus .............. | ♂ ad. | Beaver Creek, Utah... | Sept. 24 | Y. & H. |
| 301 | ...... do ..................... | ♀ ad. | ....do ................. | Sept. 24 | Y. & H. |
| 472 | ...... do ..................... | ♀ jun. | Provo, Utah .......... | Nov. 30 | Y. & H. |
| 438 | Archibuteo sancti-johannis . | ♂ | ....do ................. | Nov. 26 | Y. & H. |
| 446 | ...... do ..................... | ♂ | ....do ................. | Nov. 27 | Y. & H. |
| 447 | ...... do ..................... | ♂ | ....do ................. | Nov. 27 | Y. & H. |
| 448 | ...... do ..................... | ♀ jun. | ....do ................. | Nov. 27 | Y. & H. |
| 458 | ...... do ..................... | ♂ ad. | ....do ................. | Nov. 30 | Y. & H. |
| 466 | ...... do ..................... | ♂ ad. | ....do ................. | Nov. 30 | Y. & H. |
| 429 | ...... do ..................... | ♂ ad. | ....do ................. | Nov. 25 | Y. & H. |
| 437 | ...... do ..................... | ♂ | ....do ................. | Nov. 26 | Y. & H. |
| 485 | ...... do ..................... | ♀ ad. | ....do ................. | Dec. 1 | Y. & H. |
| 488 | ...... do ..................... | ♂ | ....do ................. | Dec. 2 | Y. & H. |
| 491 | ...... do ..................... | ♂ | ....do ................. | Dec. 3 | Y. & H. |
| 439 | Circus cyaneus, var. hudsonius. | ♂ ad. | Beaver, Utah ......... | Nov. 26 | H. & Y. |
| 440 | ...... do ..................... | ♂ jun. | ....do ................. | Nov. 26 | H. & Y. |
| 470 | ...... do ..................... | ♀ ad. | Provo, Utah .......... | Nov. 30 | H. & Y. |
| 482 | ...... do ..................... | ♂ jun. | ....do ................. | Dec. 1 | H. & Y. |
| 489 | ...... do ..................... | ♂ jun. | ....do ................. | Dec. 2 | H. & Y. |

CATHARTIDÆ.

*Rhinogryphus aura*, (L.)—Turkey-Buzzard.
Very common throughout Nevada and Utah, but extremely shy.

*Pseudogryphus californianus*, (Shaw.)—California Vulture.
A very large vulture seen near Beaver November 25 was believed to be of this species; in company with the *R. aura*, which it greatly exceeded in size. It had just finished a repast upon the carcass of a horse.

## COLUMBIDÆ.

*Zenaidura carolinensis*, (L.)—Common Dove.

Common everywhere on the plains; occurs sparingly in mountains. A number of nests were found near Provo, some containing young fully fledged July 30, and others at this time contained eggs; while in other cases the nests were still in process of construction.

| No. | Name. | Sex. | Locality. | Date. | Collector. |
|---|---|---|---|---|---|
| 29 | Zenaidura carolinensis | ♂ ad. | Provo, Utah | Aug. 2 | H. & Y. |
| 37 | ...... do | ♂ ad. | ....do | July 30 | H. & Y. |
| 47 | ...... do | ♀ ad. | ....do | July 25 | H. & Y. |
| 71 | ...... do | ♀ ad. | ....do | July 26 | H. & Y. |
| 302 | ...... do | ♀ jun. | Toquerville, Utah | Oct. 15 | H. |

## TETRAONIDÆ.

*Canace obscurua*, Say.—Dusky Grouse.

Very common on the mountains, and singularly unsuspicious and stupid; often allowing an approach close enough to strike them with a stick.

*Centrocercus urophasianus,* (Bp.)—Sage-Cock.

Numerous on plains and in mountain-valleys about 8,000 feet above water-level.

*Pediœcetes phasianellus,* (Linn.), var. *columbianus* Ord.—Sharp-tailed Grouse.

A single band seen about the middle of September in grassy foot-hills near Meadow Creek, Utah.

| No. | Name. | Sex. | Locality. | Date. | Collector. |
|---|---|---|---|---|---|
| 61 | Canace obscurua | ♀ ad. | Hobble Creek Cañon.. | Aug. 9 | H. |
| 69 | ...... do | Jun. | Danville Cañon | Aug. 12 | H. |
| E | ...... do | Jun. | ....do | Aug. 12 | H. |

## PERDICIDÆ.

*Ortyx virginianus*, (L.)—Quail.

A number of pairs of this bird were introduced at Provo from the East a few years since, and everything would seem to indicate their rapid increase. In July the call-notes of the males were frequently heard, and a number of bevies were seen here in the fall near the thickets and hedges. They are carefully protected by law, a heavy fine being imposed for their destruction.

*Lophortyx gambeli*, Nutt.—Gambel's Partridge.

This beautiful species, which is different from the California quail, although called such in Utah, was first met with early in October at

Harmony, Southern Utah, in large numbers, where it is resident all the year. The young two-thirds grown were taken at this place October 9. This locality would appear to be about its northern breeding-limit, but information was received of the occasional appearance of these birds at Cedar City, some thirty miles to the northward.

From Harmony southward it was found even more abundantly, frequenting the grain-fields and vineyards about the towns, where bevies of even one hundred were not infrequent.

Being rarely disturbed, it is quite tame, and unless closely pursued seldom takes wing, preferring to trust to its speed of foot. At Harmony many bevies habitually roosted in the heavy brush along the banks of the small streams, which are conducted through the fields, resorting thither at early dusk and departing about sunrise for the rocky hills. For rocky ground it shows great preference, and when flushed in the vicinity of such invariably betakes itself thither for concealment.

| No. | Name. | Sex. | Locality. | Date. | Collector. |
|---|---|---|---|---|---|
| 266 | Lophortyx gambeli | ♂ ad. | Harmony, Utah | Oct. 9 | H. |
| 267 | ...... do | ♀ juu. | ....do | Oct. 9 | H. |
| 268 | ...... do | ♀ jun. | ....do | Oct. 10 | H. |
| 269 | ...... do | ♂ juu. | ....do | Oct. 10 | H. |
| 280 | ...... do | ♀ ad. | ....do | Oct. 11 | H. |
| 282 | ...... do | ♂ ad. | ....do | Oct. 12 | H. |
| 293 | ...... do | ♂ ad. | ....do | Oct. 14 | H. |
| 333 | ...... do | ♀ juu. | Washington, Utah | Oct. 22 | H. & Y. |
| 323 | ...... do | ♀ ad. | Toquerville, Utah | Oct. 19 | H. & Y. |
| 338 | ...... do | ♀ ad. | Washington, Utah | Oct. 22 | H. & Y. |
| 339 | ...... do | ♀ | ....do | Oct. 22 | H. & Y. |
| 337 | ...... do | ♀ | ....do | Oct. 22 | H. & Y. |
| 343 | ...... do | ♀ | ....do | Oct. 22 | H. & Y. |
| 322 | ...... do | ♀ juu. | ....do | Oct. 22 | H. & Y. |
| 334 | ...... do | ♂ ad. | ....do | Oct. 22 | H. & Y. |
| 353 | ...... do | ♂ ad. | ....do | Oct. 22 | H. & Y. |
| 341 | ...... do | ♂ ad. | ....do | Oct. 22 | H. & Y. |
| 342 | ...... do | ♂ ad. | ....do | Oct. 22 | H. & Y. |
| 335 | ...... do | ♂ ad. | ....do | Oct. 22 | H. & Y. |
| 336 | ...... do | ♂ ad. | ....do | Oct. 22 | H. & Y. |

CHARADRIIDÆ.

*Ægialitis vociferus*, (L.)—Killdeer.

Found extremely numerous in Eastern Nevada and in the vicinity of Provo, near Utah Lake.

| No. | Name. | Sex. | Locality. | Date. | Collector. |
|---|---|---|---|---|---|
| 45 | Ægialitis vociferus | ♂ ad. | Provo, Utah | July 25 | Y. & H. |
| 46 | ...... do | ♂ ad. | ....do | July 25 | Y. & H. |

SCOLOPACIDÆ.

*Gallinago wilsonii*, (Temm).—English Snipe.

Common in marshes in Eastern Nevada after the middle of August. In November and December seen at Beaver and Provo in vicinity of warm springs in numbers, and occasionally even running along the sandy shores of streams, in this respect resembling the sandpipers. Winters at least as far north as Provo.

*Macrorhamphus griseus*, (Gm.)—Red-breasted Snipe.

Observed at Provo, July 24, and probably breeds in this vicinity. Common during the fall-migration.

*Tringa minutilla*, V.—Least Sandpiper.

A single individual taken July 26; few only seen.

*Eureunetes pusillus*, (L.)—Semipalmated Sandpiper.

We are indebted for a single specimen of this bird to Mr. G. K. Gilbert, geologist of the party, who obtained it at Sevier Lake, Utah, in September, a large flock being seen at this time.

*Gambetta melanoleuca*, (Gm.)—Tell-tale Stone-Snipe or Greater Yellow-legs.

Common in Nevada and Utah during the fall-migrations.

*Tringoides macularius*, (L.)—Spotted Sandpiper.

Breeds near Provo, and is quite common in all the streams.

*Numenius longirostris*, Wils.—Long-billed Curlew.

Very numerous in sloughs near Fairfield, Utah, and tolerably common in Eastern Nevada near small lakes. A wounded specimen taken at Fillmore in November.

| No. | Name. | Sex. | Locality. | Date. | Collector. |
|---|---|---|---|---|---|
| 82 | Tringa minutilla | ♂ ad. | Utah Lake | July 26 | Y. & H. |
| 177 | Macrorhamphus griseus | ♀ ad. | .... do | July 24 | Y. & H. |
| 199 | ...... do | ♀ jnn. | Rush Lake, Utah | Oct. 1 | H. |
| 147 | Gambetta melanoleuca | ♀ jun. | Deep Creek, Utah | Aug. 12 | Y. |
| 432 | Gallinago wilsonii | ♂ ad. | Provo, Utah | Nov. 25 | Y. & H. |
| 433 | ...... do | ♂ ad. | .... do | Nov. 25 | Y. & H. |
| 24 | Tringoides macularius | ♀ jun. | .... do | July 24 | Y. & H. |
| E | Eureunetes pusillus | | Sevier Lake, Utah | Sept. — | Gilbert. |
| 427 | Numenius longirostris | ♂ ad. | Fillmore, Utah | Nov. 19 | Y. & H. |
| 174 | ...... do | | Fairfield, Utah | Aug. 3 | Y. |

PHALAROPODIDÆ.

*Steganopus wilsonii*, (Sab.)—Wilson's Phalarope.

Seen at Great Salt Lake in July.

RECURVIROSTRIDÆ.

*Recurvirostra americana*, Gm.—American Avocet.

Numerous in August at Fairfield, and in Eastern Nevada, and present at Rush Lake, Utah, October 1, in large flocks. Very shy and wary.

*Himantopus nigricollis*, V.—Black-necked Stilt.

Rather common at Fairfield, Utah, in August.

| No. | Name. | Sex. | Locality. | Date. | Collector. |
|-----|-------|------|-----------|-------|------------|
| 441 | Recurvirostra americana ... | ♂ | Provo, Utah.......... | Nov. 26 | Y. & H. |
| C7 | Himantopus nigricollis. .... | ....... | Fairfield, Utah ....... | Aug. — | Y. |

### GRUIDÆ.

*Grus canadensis*, Temm.—Sandhill-Crane.

First seen at Fish Springs, Utah, in August. Companies of two or three were afterward observed on the plains later in the season.

### TANTALIDÆ.

*Tantalus loculator*, L.—Wood-Ibis.

A flock of eight or ten individuals seen at Rush Lake in October, and two fine specimens secured.

*Ibis guarauna* (Gm.)—Glossy Ibis.

Although not met with, it is well known to gunners in the vicinity of Utah Lake as the "Black Snipe."

| No. | Name. | Sex. | Locality. | Date. | Collector. |
|-----|-------|------|-----------|-------|------------|
| 201 | Tantalus loculator.......... | ♂ ad. | Rush Lake, Utah ..... | Oct. 1 | H. |
| 202 | ...... do ..................... | ♀ ad. | .... do ............... | Oct. 1 | H. |

### ARDEIDÆ.

*Ardea herodias*, L.—Great Blue Heron.

Observed as common at Utah Lake and Rush Lake, a few being seen at the former place as late as December.

*Herodias egretta*, (Gm.)—White Heron.

A single individual observed near Beaver, Utah, but not secured. From information received it is probably not common at Provo, though seen there.

*Botaurus minor*, (Gm.)—Bittern.

Not uncommon. Two specimens secured in Southern Utah.

*Nyctiardia grisea*, (L.), var. *nævia* Bodd.—Night-Heron.

Common at Provo, where it was seen in December.

| No. | Name. | Sex. | Locality. | Date. | Collector. |
|-----|-------|------|-----------|-------|------------|
| 167 | Botaurus minor ............. | ♀ ad. | Panquitch, Utah...... | Sept. 17 | H. |
| 300 | ...... do ..................... | ♂ ad. | Beaver, Utah ......... | Sept. 24 | Y. & H. |
| 168 | Nyctiardia grisea, var. nævia ................. | ♀ jun. | .... do............... | Sept. 24 | Y. & H. |

## RALLIDÆ.

*Porzana carolina*, (L.)—Sora Rail.

*Rallus virginianus*, L.—Virginia Rail.
Common at Provo, the latter being taken in November.

*Fulica americana*, Gm.—Coot.
Very numerous at Rush Lake in September.

| No. | Name. | Sex. | Locality. | Date. | Collector. |
|---|---|---|---|---|---|
| 83 | Porzana carolina | ♀ jun. | Utah Lake, Utah | July 26 | Y. & H. |
| 454 | Rallus virginianus | ♂ ad. | Provo, Utah | Nov. 27 | Y. & H. |
| 109 | Fulica americana | ♂ | Panquitch, Utah | Sept. 17 | H. |

## ANATIDÆ.

*Anser hyperboreus*, Pall.—Snow-Goose.
Immense gangs of this bird noticed at Rush Lake, Utah, in early November.

*Branta canadensis*, (L.)—Canada Goose.
Extremely numerous during fall and early winter in all the lakes in Utah ; it passes the nights in the water, returning to the grain-fields at daybreak to feed.

*Branta bernicla*, (L.), var. *nigricans* Lawr.—Black Brant.
Brant were seen at Rush Lake, supposed to be of this species.

*Anas boschas*, L.—Mallard.
One of the most numerous of all the ducks in Nevada and Utah, wintering near warm springs. Young scarcely able to fly by the middle of August.

*Anas obscura*, Gm.—Black Duck.
Seen only at Rush Lake in November (Yarrow).

*Dafila acuta* (L.)—Sprig- or Pin-tail Duck.
Common.

*Nettion carolinensis*, (Gm.)—Green-winged Teal.
Common throughout Nevada and Utah.

*Querquedula discors*, (L.), Steph.—Blue-winged Teal.
Not nearly as common as preceding.

*Querquedula cyanoptera*, (V.)—Red-breasted Teal.
Breeds in great numbers in the marshes of Utah Lake, migrating south very early. None taken.

*Spatula clypeata*, (L.)—Shoveler.
Very common.

*Chaulelasmus streperus*, (L.)—Gadwall.

Few only seen late in November at Provo.

*Marcca americana*, (Gm.)—Baldpate.

Numerous.

*Fuligula marila*, (L.)—Greater Blackhead.

Rather common.

*Fuligula collaris*, (Donov.)—Ring-necked Duck.

Single young female taken at Rush Lake in September.    Probably migrates south comparatively early in the season.

*Fuligula ferina*, (L.), var. *americana* Eyton.—Redhead.

But few seen.

*Bucephala clangula*, (L.)—Golden Eye.

Numerous in Provo River.

*Bucephala islandica*, (Gm.)—Barrow's Golden Eye.

A pair of these ducks were taken in the Provo River December 1. This is the first instance of the discovery of this bird so far inland, and to the southward. Believed not to be uncommon, as gunners distinguish it from the preceding by its large size.

*Bucephala albeola*, (L.)—Butter-Ball.

Common.

*Erismatura rubida*, (Wils.)—Ruddy Duck.

Rather uncommon; taken at Provo in November.

*Mergus serrator*, L.—Red-breasted Merganser.

Very common.

*Mergus cucullatus*, L.—Hooded Merganser.

Rather common.

NOTE.—The observations upon the foregoing species were made in Spring and Snake Valleys, Nevada, and Rush Lake and Utah Lake, Utah; but as more time was spent at Utah Lake than at the other points, it may be assumed that the notes apply more particularly to this locality.

| No. | Name. | Sex. | Locality. | Date. | Collector. |
|---|---|---|---|---|---|
| 171 | Anas boschas | ♀ ad. | Fairfield, Utah | Aug.  3 | Y. |
| 124 | Nettion carolinensis | ♀ ad. | Thistle Valley, Utah. | Aug. 23 | H. |
| 193 | ...... do | ♂ jun. | Deep Creek, Utah | Aug. 12 | Y. |
| 452 | ...... do | ♂ ad. | Provo, Utah | Nov. 27 | H. & Y. |
| 109 | Querquedula discors | ♀ ad. | Thistle Valley, Utah. | Aug. 18 | H. |
| F163 | ...... do | ♂ | Fairfield, Utah | Aug.  3 | Y. |
| 197 | Spatula clypeata | ♀ | Rush Lake, Utah | Sept. 30 | H. |
| 198 | ...... do | ♀ ad. | ....do | Oct.  2 | H. & S. |
| 450 | Mareca americana | ♂ ad. | Provo, Utah | Nov. 27 | H. & S. |
| 194 | Fuligula ferina, var. americana. | ♂ jun. | ....do | Nov. 27 | H. |
| 203 | ...... do | ♂ jun. | Rush Lake, Utah | Oct.  1 | H. & S. |
| 430 | Bucephala islandica | ♂ ad. | Provo, Utah | Nov. 11 | H. & S. |
| 483 | ...... do | ♀ ad. | ....do | Dec.  1 | H. & S. |
| 490 | Bucephala clangula | ♂ ad. | ....do | Nov.  2 | H. & S. |
| 431 | Bucephala albeola | ♀ jun. | ....do | Nov. 25 | H. & S. |
| 457 | ...... do | ♂ ad. | ....do | Nov. 25 | H. & S. |
| 203 | Fuligula collaris | ♂ jun. | Rush Lake, Utah | Oct.  1 | H. |
| 453 | Erismatura rubida | ♂ jun. | Provo, Utah | Nov. 27 | Y. & H. |
| 449 | Chaulelasmus streperus | ♂ jun. | ....do | Nov. 27 | Y. & H. |

## PELECANIDÆ.

*Pelecanus erythrorhynchus*, Gm.—American Pelican.

In Stansbury's report of Great Salt Lake mention is made of large numbers of these birds being seen in the lake, they breeding in the islands thereof. In July but few were seen, and we are informed they no longer breed there. These birds were seen at Utah Lake late in July sparingly, and in September on the sloughs of the Sevier.

## LARIDÆ.

*Larus delawarensis*, Ord.—Ring-billed Gull.

Common on the lakes throughout Utah. Numbers seen on the Provo River late in November when the lake was frozen.

*Sterna forsteri*, Nutt.—Havell's Tern.

Common at Utah Lake, where it only was seen.

*Hydrochelidon fissipes*, Gr.—Short-tailed Tern.

Only two individuals seen, at Utah Lake in July.

NOTE.—Stansbury also mentions the occurrence of numerous gulls in Salt Lake; these of late years have greatly decreased. Several gulls were seen but not identified.

| No. | Name. | Sex. | Locality. | Date. | Collector. |
|---|---|---|---|---|---|
| 467 | Larus delawarensis......... | ♀ | Provo, Utah .......... | Nov. 30 | Y. & H. |
| 12 | Sterna forsteri .............. | ♀ ad. | ....do ............... | July 24 | Y. & H. |
| 13 | ...... do ................... | ♀ ad. | ....do ............... | July 24 | Y. & H. |
| 14 | ...... do ................... | ♂ ad. | ....do ............... | July 24 | Y. & H. |
| 15 | ...... do ................... | ♂ ad. | ....do ............... | July 24 | Y. & H. |
| 16 | ...... do ................... | ♂ ad. | ....do ............... | July 24 | Y. & H. |
| 17 | ...... do ................... | ♂ ad. | ....do ............... | July 24 | Y. & H. |
| 18 | ...... do ................... | ♀ ad. | ....do ............... | July 24 | Y. & H. |
| 19 | ...... do ................... | ♀ ad. | ....do ............... | July 24 | Y. & H. |

## COLYMBIDÆ.

*Colymbus torquatus*, Brunn.—Great Northern Diver.

Said to be rather common at Utah Lake.

## PODICIPIDÆ.

*Podiceps occidentalis*, Lawr.—Western Grebe.

Common on Utah Lake. One specimen secured.
*Podiceps cornutus*, Lath.—Horned Grebe.
*Podilymbus podiceps*, Lawr.—Carolina Grebe.

Few of the former and many of the latter seen at Rush Lake in September.

| No. | Name. | Sex. | Locality. | Date. | Collector. |
|---|---|---|---|---|---|
| 31 | Podiceps occidentalis ....... | Ad. | Utah Lake, Utah ..... | July 24 | H. & S. |

## LIST OF BIRDS COLLECTED BY LIEUT. G. M. WHEELER'S EXPEDITION, 1871.

The following list of birds, collected in 1871 by Mr. Bischoff and other members of the party, will be found to represent some species whose geographical distribution has been greatly increased in the past two years. There are four species new to the fauna of Nevada; they are *Campylorhynchus brunneicapillus, Phainopepla nitens, Guiraca cœrulea,* and *Tantalus loculator.* The specimens were nearly all secured in Arizona and Nevada, and it is greatly to be regretted that fuller notes cannot be given, the MSS. of Mr. Bischoff having been destroyed by fire. When we take into consideration the extreme barrenness of the localities visited, and the difficulties under which the collections were made, we cannot help thinking that a great deal has been accomplished; for, notwithstanding the small number of specimens secured, 88 in all, no less than 64 species, many of them rare, are represented in the collection. What with the ornithological labors of Dr. Coues in Arizona, those of Lieutenant Wheeler in the same Territory, Nevada, and Utah, and of Messrs. Allen and Ridgway in the latter Territory, American ornithology has certainly received a new impetus and acquired many valuable facts heretofore unknown.

*List of birds collected by Lieutenant Wheeler's expedition in 1871.*

| Name. | Sex. | Locality. | Date. | Collector. |
|---|---|---|---|---|
| TURDIDÆ. | | | | |
| Oreoscoptes montanus | | Nevada | June 20 | Bischoff. |
| Mimus caudatus | ♀ ad. | Arizona | Nov. 30 | Do. |
| SAXICOLIDÆ. | | | | |
| Sialia arctica | ♂ ad. | Nevada | June 20 | Do. |
| | | | | Do. |
| SITTIDÆ. | | | | Do. |
| | | | | Do. |
| Sitta pygmæa | ♂ ad. | Arizona | Nov. 14 | Do. |
| Do | ♂ ad. | ....do | Nov. 14 | Do. |
| Do | ♂ ad. | ....do | Nov. 14 | Do. |
| TROGLODYTIDÆ. | | | | |
| Campylorhynchus brunneicapillus | Ad. | Nevada | Aug. 14 | Do. |
| SYLVICOLIDÆ. | | | | |
| Dendroica æstiva | ♀ ad. | ....do | Aug. 9 | Do. |
| AMPELIDÆ. | | | | |
| Phainopepla nitens | Jun. | ....do | Aug. 6 | Do. |
| LANIIDÆ. | | | | |
| Collurio excubitoroides | Ad. | ....do | June 1 | Do. |
| Do | Ad. | ....do | Sept. 19 | Do. |
| FRINGILLIDÆ. | | | | |
| Passerculus alaudinus | ♂ ad. | ....do | Sept. 8 | Do. |
| Pocœcetes confinis | ♀ ad. | ....do | June 20 | Do. |
| Coturniculus perpallidus | ♂ ad. | ....do | Sept. 8 | Do. |

List of birds collected by Lieutenant Wheeler's expedition in 1871—Continued.

| Name. | Sex. | Locality. | Date. | Collector. |
|---|---|---|---|---|
| Melospiza hermanni | Ad. | Arizona | Sept. 8 | Do. |
| Spizella breweri | Ad. | ...do | Sept. 8 | Do. |
| Zonotrichia leucophrys | ♂ ad. | Bull Run, Nevada. | May 25 | Do. |
| Zonotrichia intermedia | Ad. | Nevada | Oct. 30 | Do. |
| Poospiza bilineata ..: | ♂ ad. | ...do | | Do. |
| Do | Jun. | ...do | | Do. |
| Hedymeles melanocephalus | ♂ ad. | Bull Run, Nevada. | May 25 | Do. |
| Guiraca cœrulea | ♂ ad. | Nevada | Aug. 9 | Do. |
| Cyanospiza amœna | ♂ ad. | Bull Run, Nevada. | May 25 | Hoffman. |
| Do.. | ♂ ad. | ...do | May 25 | Bischoff. |
| Do | ♂ ad. | ...do | May 23 | Hoffman. |
| Cardinalis igneus | ♂ ad. | Arizona | Nov. 30 | Bischoff. |
| Pyrrhuloxia sinuata | ♂ ad. | ...do ...: | Nov. 30 | Do. |
| Pipilo chlorurus | ♂ ad. | Bull Run, Nevada. | May 23 | Hoffman. |
| Do...: | ♀ ad. | ...do | May 24 | Do. |
| Do | ♀ ad. | ...do | June 24 | Bischoff. |
| ALAUDIDÆ. | | | | |
| Eremophila alpestris | ♂ ad. | Nevada | June 10 | Do. |
| ICTERIDÆ. | | | | |
| Icterus bullockii | ♂ ad. | ...do | June 4 | Do. |
| Xanthocephalus icterocephalus | ♂ ad. | Halleck, Nevada .. | May 14 | Do. |
| Do | ♂ ad. | ...do | May 14 | Do. |
| Do | ♂ ad. | ...do | May 14 | Do. |
| Do | ♂ ad. | ...do | May 14 | Do. |
| Do | ♂ ad. | ...do | May 14 | Do. |
| Do | ♂ ad. | ...do | May 14 | Do. |
| Do | ♂ ad. | ...do | May 14 | Do. |
| Do | ♂ ad. | ...do | May 14 | Do. |
| Do | ♂ ad. | ...do | May 14 | Do. |
| Do | ♂ ad. | ...do | May 14 | Do. |
| Do | ♂ ad. | ...do | May 14 | Do. |
| Do | ♀ ad. | ...do | May 14 | Do. |
| Do | Jun. | ...do | May 14 | Do. |
| CORVIDÆ. | | | | |
| Cyanura macrolopha | ♂ ad. | Arizona | Nov. 20 | Do. |
| Do | ♂ ad. | ...do | Nov. 20 | Do. |
| Cyanocitta woodhousei | ♀ ad. | Nevada | Sept. 6 | Do. |
| Picicorvus columbianus | Ad. | ...do | July — | Do. |
| Gymnokitta cyanocephala | Jun. | ...do | Sept. 8 | Do. |
| TYRANNIDÆ. | | | | |
| Tyrannus verticalis | ♀ ad. | ...do | May 25 | Do. |
| Do | Ad. | ...do | May 25 | Do. |
| Do | Jun. | ...do | Sept. 14 | Do. |
| Do | ♂ ad. | California | Aug. 16 | Do. |
| Contopus borealis | ♂ jun. | Nevada | Aug. 16 | Do. |
| Empidonax pusillus | ♀ ad. | Humboldt River, Nevada. | May 31 | Do. |
| Empidonax obscurus | ♀ ad. | Bull Run, Nevada. | May 25 | Hoffman. |
| Do | Ad. | Nevada | | Bischoff. |
| CAPRIMULGIDÆ. | | | | |
| Chordeiles henryi | ♀ ad. | ...do | Aug. 9 | Do. |
| TROCHILIDÆ. | | | | |
| Stellula calliope | ♀ ad. | ...do | May 25 | Do. |

*List of birds collected by Lieutenant Wheeler's expedition in 1871—Continued.*

| Name. | Sex. | Locality. | Date. | Collector |
|---|---|---|---|---|
| **PICIDÆ.** | | | | |
| Sphyropicus nuchalis ............... | ♂ ad. | Arizona .......... | Oct. 31 | Do. |
| Melanerpes formicivorus ............ | ♀ ad. | ....do ............ | Nov. 20 | Do. |
| **STRIGIDÆ.** | | | | |
| Otus wilsonianus .................... | ♂ ad. | Carlin, Nevada ... | May 19 | Hoffman. |
| **FALCONIDÆ.** | | | | |
| Buteo calurus....................... | ........ | Arizona .......... | Nov. 6 | Bischoff. |
| Do........................ | ........ | ....do .......... | Nov. 6 | Do. |
| Buteo swainsoni..................... | ♂ ad. | Antelope, Nevada . | May 28 | Do. |
| Falco polyagrus .................... | ♀ jun. | Arizona .......... | Nov. 3 | Do. |
| Circus hudsonius ................... | ♀ jun. | Nevada .... ...... | Sept. 10 | Do. |
| Do........................ | ♀ ad. | ....do ............ | Sept. 10 | Do. |
| Nisus cooperi...................... | Jun. | ....do .......... | Sept. 6 | Do. |
| Falco sparverius.................... | ♂ ad. | Arizona .......... | Oct. 30 | Do. |
| Do........................ | ♀ jun. | Nevada .... ...... | Sept. 8 | Do. |
| **MELEAGRIDIDÆ.** | | | | |
| Meleagris mexicana.................. | Ad. | Arizona ..... .... | ......... | Do. |
| **TETRAONIDÆ.** | | | | |
| Centrocercus urophasianus .......... | ♂ ad. | Nevada ..... .... | ......... | Do. |
| **PERDICIDÆ.** | | | | |
| Lophortyx gambelii ............... | ♀ ad. | Arizona .......... | Sept. 8 | Do. |
| Cyrtonyx massena ................... | ♂ ad. | ....do ............ | Nov. 18 | Do. |
| **SCOLOPACIDÆ.** | | | | |
| Symphemia semipalmata............ | Ad. | .................... | May 5 | R. |
| **RECURVIROSTRIDÆ.** | | | | |
| Himantopus nigricollis......... .-... | ♀ ad. | .................... | Sept. 3 | Bischoff. |
| **TANTALIDÆ.** | | | | |
| Tantalus loculator.................. | ......... | Nevada .... ...... | July — | Do. |
| **ARDEIDÆ.** | | | | |
| Nyctiardea nævia................... | Ad. | (?) .............. | ......... | (?) |
| Do.......................... | Ad. | (?) .............. | ......... | (?) |
| Botaurus minor .......... ......... | Ad. | .................... | May — | Hoffman. |
| **RALLIDÆ.** | | | | |
| Fulica americana .................. | Ad. | (?) .............. | June 4 | Bischoff. |
| Do.......................... | Ad. | (?) .............. | May 4 | Do. |
| Do.......................... | Jun. | (?) .............. | May 11 | R. |
| **PODICIPIDÆ.** | | | | |
| Podiceps californicus............... | ♂ jun. | Nevada .... ...... | Sept. 12 | R. |

*List of alcoholic specimens of birds.*

| No. | Name. | Collector. |
|---|---|---|
| | TURDIDÆ. | |
| A1 | Turdus migratorius............................................... | H. & Y. |
| A2 | Oreoscoptes montanus ..................... ............ | H. & Y. |
| | PARIDÆ. | |
| A3 | Parus montanus....................:............................ | H. & Y. |
| A4 | ......do ...........................•............................... | H. & Y. |
| A5 | Psaltriparus plumbeus........................................ | H. & Y. |
| A6 | ......do ........................................................ | H. & Y. |
| | TROGLODYTIDÆ. | |
| A7 | Salpinctes obsoletus (skull) ................................. | H. & Y. |
| A8 | ......do...........................do.......................... | H. & Y. |
| | MOTACILLIDÆ. | |
| A9 | Anthus ludovicianus ........................................ | H. & Y. |
| | SYLVICOLIDÆ. | |
| A10 | Dendroica æstiva............................................ | H. & Y. |
| A11 | ......do ....................................................... | H. & Y. |
| A12 | Geothlypis trichas............................................ | H. & Y. |
| A13 | Geothlypis macgillivrayi ..................................... | H. & Y. |
| | HIRUNDINIDÆ. | |
| A14 | Stelgidopteryx serripennis ......•......................... | H. & Y. |
| | LANIIDÆ. | |
| A15 | Collurio ludovicianus, var. excubitoroides .................... | H. & Y. |
| | FRINGILLIDÆ. | |
| A16 | Chrysomitris tristis........................................... | H. & Y. |
| A17 | ......do ....................................................... | H. & Y. |
| A18 | ......do ....................................................... | H. & Y. |
| A19 | ......do ....................................................... | H. & Y. |
| A20 | Carpodacus frontalis.......................................... | H. & Y. |
| B1 | Cyanospiza amœna ........................................... | H. & Y. |
| B2 | Spizella pallida, var. breweri................................ | H. & Y. |
| B3 | ......do ....................................................... | Y. & H. |
| B4 | Spizella monticola............................................ | Y. & H. |
| B5 | Melospiza melodia, var. fallax................................ | Y. & H. |
| B6 | Zonotrichia leucophrys, var. intermedia....................... | Y. & H. |
| B7 | ......do ....................................................... | Y. & H. |
| B8 | Poocætes gramineus.......................................... | Y. & H. |
| B9 | Pipilo megalonyx ........................................... | Y. & H. |
| B10 | ......do ....................................................... | Y. & H. |
| | ALAUDIDÆ. | |
| B11 | Eremophila alpestris.......................................... | Y. & H. |
| | ICTERIDÆ. | |
| B12 | Sturnella magna, var. neglecta ............................... | Y. & H. |
| B13 | ......do ....................................................... | Y. & H. |
| B14 | Dolichonyx oryzivorus........................................ | Y. & H. |

*List of alcoholic specimens of birds.*

| No. | Name. | Collector. |
| --- | --- | --- |
| B15 | Xanthocephalus icterocephalus | Y. & H. |
| B16 | Scolecophagus cyanocephalus | Y. & H. |
| | **TYRANNIDÆ.** | |
| B17 | Tyrannus verticalis | Y. & H. |
| B18 | Empidonax pusillus | Y. & H. |
| | **ALCEDINIDÆ.** | |
| B19 | Ceryle alcyon | Y. & H. |
| | **STRIGIDÆ.** | |
| B20 | Otus vulgaris, var. wilsonianus | H. & Y. |
| C1 | ......do | Y. & H. |
| | **FALCONIDÆ.** | |
| C2 | Falco sparverius | Y. & H. |
| | **PERDICIDÆ.** | |
| C3 | Lophortyx gambelii (skulls) | Y. & H. |
| C4 | ......do | Y. & H. |
| | **CHARADRIIDÆ.** | |
| C5 | Ægialitis vociferus | Y. & H. |
| C6 | ......do | Y. & H. |
| | **RECURVIROSTRIDÆ.** | |
| C7 | Himantopus nigricollis (skull) | Y. & H. |
| C8 | Recurvirostra americana (skull) | Y. & H. |
| | **ANATIDÆ.** | |
| C9 | Mergus serrator (skull) | Y. & H. |
| C10 | Querquedula cyanoptera (skull) | Y. & H. |
| | **LARIDÆ.** | |
| C11 | Larus delawarensis (skull) | Y. & H. |
| C12 | ......do | Y. & H. |
| | **PODICIPIDÆ.** | |
| C13 | Podilymbus podiceps | Y. & H. |

NOTE.—The above specimens are those of birds too badly mutilated to make skins of, and those where time would not admit of their proper preservation. In addition, a large collection of *sterna* and tongues of birds in alcohol was made.

# I.—AN ANNOTATED LIST OF THE BIRDS OF UTAH.*

## BY H. W. HENSHAW.

The following list is based largely upon material collected during the field-season of 1872, while with the exploring and surveying party in charge of Lieut. G. M. Wheeler, of the United States Engineers. In it are enumerated all the birds thus far known to have been taken or observed within the limits of the territory. To give it additional value as a formal list, those known to breed, whether from actual observations in the field or from their known breeding-range, are indicated.† Notes are also given respecting their relative abundance or scarcity. Of the 214 species given, 160 were either actually taken or noted by Dr. Yarrow and myself during the season. Of the remaining species, 25 not met with by us are contained in Mr. Allen's list of birds, collected in the vicinity of. Ogden, from September 1 to October 8. I am also indebted to Mr. Ridgway for a list of the birds noted by him during his collecting trip in this locality, including many not contained in either Mr. Allen's paper or our own report, and also for assistance in the preparation of the list. It may be here stated that no collections have ever been made in Utah during the spring-months, and thus many of the spring-migrants have entirely escaped notice. This will account for the comparatively small number of species mentioned. An entire season's connected observations would doubtless add many to the number.

## TURDIDÆ (the Thrushes).

*1. *Turdus fuscescens*, Steph.—Tawny Thrush.

Summer-resident. Common on Provo River in summer of 1869. (Ridgway.)

*2. *Turdus swainsoni*, Cab.—Olive-backed Thrush.

Very common. Inhabits the thickets of the mountain-streams. (Ridgway.)

*3. *Turdus pallasi*, Cab., var. *audubonii*, Bd.—Rocky-Mountain Hermit-Thrush.

Less common than the preceding. Inhabits the pine-region. (Ridgway.) Ogden, September. (Allen.)

*4. *Turdus migratorius*, L.—Robin.

Very common. Permanent resident.

---

* This paper was read April 6, 1874, before the Lyceum of Natural History, New York, and printed in Annals of the Lyceum, vol. xi, June, 1874. In its present form it is substantially the same, with the exception of revisions made to accord with the present state of knowledge upon the subject.

† An asterisk (*) is prefixed to the names of those known to breed in the Territory.

*5. *Oreoscoptes montanus* (Towns.)—Mountain Mocking-Bird.
An inhabitant of the valleys and plains. Most abundant in the neighborhood of settlements.

*6. *Harporhynchus crissalis*, Henry.—Red-vented Thrush.
Resident (?). Found breeding, and nest and eggs obtained by Dr. Palmer at Saint George. Seen by me in same locality.

*7. *Galeoscoptes carolinensis* (L.)—Catbird.
Very abundant. Inhabits the thickets.

CINCLIDÆ (the Water-Ouzels).

*8. *Cinclus mexicanus*, Swains.—Water-Ouzel; Dipper.
Very abundant. Inhabiting the rapid mountain-streams. Permanent resident.

SAXICOLIDÆ (the Stone-Chats).

*9. *Sialia arctica*, Swains.—Rocky-Mountain Bluebird.
Resident. Very abundant. "Found breeding at Salt Lake City and Antelope Island in May and June." (Ridgway.)

SYLVIIDÆ (the Sylvias).

*10. *Regulus calendula* (L.)—Ruby-crowned Kinglet.
Common resident. Found breeding high up in the mountains by Mr. Ridgway. Winters in the valleys.

PARIDÆ (the Titmice).

*11. *Lophophanes inornatus* (Gamb.)
Very abundant. Resident. Breeds in mountains. Wintering in the cedars of the valleys.

*12. *Parus montanus*, Gamb.—Mountain-Chickadee.
Abundant. Resident in the mountains.

*13. *Parus atricapillus*, L., var. *septentrionalis*, Harris.—Long-tailed Chickadee.
Abundant and resident in vicinity of Provo. Apparently not very generally distributed. Not found in mountains.

*14. *Psaltriparus minimus* (Towns.) var. *plumbeus*, Bd.—Lead-colored Tit.
Abundant, moving in large companies. Breeds in the mountains, and winters in the valleys.

SITTIDÆ (the Nuthatches).

*15. *Sitta carolinensis*, Gm., var. *aculeata*, Cass.—Slender-billed Nuthatch.
Apparently not common in the mountains. Met with on but one occasion by us. Resident.

*16. *Sitta pusilla* (Lath.) var. *pygmæa*, Vig.—Pigmy Nuthatch.
Same as preceding.

CERTHIIDÆ (the Creepers).

*17. *Certhia familiaris*, L., var. *americana*, Bon.—Brown Creeper.
Rare in the pines of mountains in June. Probably breeds. (Ridgway.)

TROGLODYTIDÆ (the Wrens).

*18. *Campylorhynchus brunneicapillus* (Lafr.)—Cactus-Wren.
Rare in southern parts of State. Several individuals seen in vicinity of Saint George, October 27. Possibly breeds.

*19. *Salpinctes obsoletus* (Say).—Rock-Wren.
Exceedingly abundant in rocky localities.

*20. *Catherpes mexicanus* (Sw.), var. *conspersus*, Ridgw.—White-throated Rock-Wren.
Rather rare, but generally distributed. Permanent resident.

*21. *Thryothorus bewickii* (Aud.), var. *leucogaster*, Gould.—Bewick's Wren.
Not uncommon in southern part of Territory in fall. Probably breeds.

*22. *Troglodytes aëdon*, Vieill., var. *parkmanni*, Aud.—Parkman's Wren.
Abundant in the mountains. Permanent resident.

*23. *Cistothorus stellaris* (Licht.) Short-billed Marsh-Wren.
Probably rare. Not taken, but evidence obtained of its breeding on borders of Utah Lake.

*24. *Telmalodytes palustris* (Wils.), var. *paludicola*, Bd.—Long-billed Marsh-Wren.
Exceedingly abundant in the marshes everywhere. Permanent resident.

MOTACILLIDÆ (the Wagtails).

25. *Anthus ludovicianus* (Gmel.)—Tit-Lark.
Abundant in the marshes. Winter-resident.

SYLVICOLIDÆ (the Warblers).

26. *Helminthophaga ruficapilla* (Wils.)—Nashville Warbler.
Apparently common. Ogden, September. (Allen.)

*27. *Helminthophaga virginiæ*, Bd.—Virginia's Warbler.
Frequent among the scrub-oaks of foot-hills, breeding. (Ridgway.)

*28. *Helminthophaga celata* (Say.)—Orange-crowned Warbler.
Breeds in mountains from 7,000 to 9,000 feet high. (Ridgway.) Common in September. Ogden. (Allen.)

*29. *Dendroica æstiva* (Gm.)—Yellow Warbler.
Very common in neighborhood of settlements.

*30. *Dendroica audubonii* (Towns.)—Audubon's Warbler.
Abundant, especially in fall. "Breeds in the pine-region of the Wahsatch." (Ridgway.)

31. *Dendroica blackburniæ* (Gm.)—Blackburnian Warbler.
"Not common. Ogden. September." (Allen.)

32(?). *Dendroica nigrescens* (Towns.)—Black-throated Gray Warbler.
Ogden. September. (Allen.)

*33. *Geothlypis philadelphia* (Wils.), var. *macgillivrayi*, Aud.—Macgilli-vray's Warbler.
Common in the mountains.

*34. *Geothlypis trichas* (L.)—Maryland Yellowthroat.
Common. Distributed generally through the valleys of the Territory, in the neighborhood of water.

*35. *Icteria virens* (L.), var. *longicauda*, Lawr.—Long-tailed Chat.
Common. Inhabiting indifferently the thickets of foot-hills and valleys.

36. *Myiodioctes pusillus* (Wils.)—Wilson's Blackcap.
Common as a spring and autumn migrant.

*37. *Setophaga ruticilla* (L.)—Redstart.
Rather common as an inhabitant of the mountains and valleys.

HIRUNDINIDÆ (the Swallows).

*38. *Progne subis* (L.)—Purple Martin.
Quite abundant in the mountains, frequenting aspen-groves.

*39. *Petrochelidon lunifrons* (Say).—Cliff-Swallow.
Very abundant in the mountains. Breeds in large colonies on the cliffs.

*40. *Hirundo horreorum* Bart.—Barn-Swallow.
Common. Builds in barns, deserted shanties, and caves.

*41. *Tachycineta bicolor* (Vieill.)—White-bellied Swallow.
Abundant. Generally distributed. Breeds in the aspen-groves in company with the martins.

*42. *Tachycineta thalassina* (Swains.)—Violet-green Swallow.
Somewhat rare. Breeds in limestone-cliffs with the White-throated Swift (*Panyptila saxatilis*) and the Cliff-Swallow. (Ridgway.)

*43. *Cotyle riparia* (L.)—Bank-Swallow.
Quite common at Provo. Breeds in the river-banks in company with the Rough-winged Swallow (*Stelgidopteryx serripennis*).

44. *Stelgidopteryx serripennis* (Aud.)—Rough-winged Swallow.
Far more numerous than the preceding, with which it is associated.

VIREONIDÆ (the Greenlets).

45. *Vireo olivaceus* (L.)—Red-eyed Vireo.
Quite common at Ogden in September. (Allen.)

*46. *Vireo gilvus* (Vieill.), var. *swainsoni*, Bd.—Warbling Viero.
Very abundant. Generally distributed. Found breeding, by Mr. Ridgway, from lowest valleys to altitude of 8,000 feet.

*47. *Vireo solitarius* (Wils.), var. *plumbeus*, Cs.—Solitary Vireo.
Rather rare.

AMPELIDÆ (the Waxwings).

48. *Ampelis cedrorum* (Vieill.)—The Cedar-Bird.
Rather common. Ogden, September. (Allen.)

MYIADESTIDÆ (the Solitaires).

*49. *Myiadestes townsendii* (Aud.)—Townsend's Solitaire.
Rather rare. Breeds on the mountains and winters in the cedar-groves of valleys.

LANIIDÆ (the Shrikes, or Butcher-Birds).

50. *Collurio borealis* (Vieill.)—Great Northern Shrike.
Of frequent occurrence in fall. Winter-resident.

*51. *Collurio ludovicianus* (L.), var. *excubitoroides*, Swains.—White-rumped Shrike.
Quite common. Permanent resident.

TANAGRIDÆ (the Tanagers).

*52. *Pyranga ludoviciana* (Wils.)—Louisiana Tanager.
Common.

FRINGILLIDÆ (the Finches, Sparrows, Buntings, &c.)

*53. *Carpodacus frontalis* (Say).—House-Finch.
Very abundant. Breeds in the valleys.

*54. *Carpodacus cassini* Bd.—Cassin's Purple Finch.
Abundant. Breeds on the mountains. (Ridgway.)

*55. *Chrysomitris pinus* (Wils.)—Pine-Finch.
Breeds abundantly in pine-regions of mountains. Resident. (Ridgway.)

*56. *Chrysomitris tristis* (L.)—Yellowbird.
Common. Permanent resident.

*57. *Chrysomitris psaltria* (Say).—Arkansas Finch.
Quite common in southern part of Territory late in fall. "Breeds sparingly near Salt Lake City and to the eastward." (Ridgway.)

58. *Leucosticte tephrocotis*, Swains.—Gray-crowned Finch.
Obtained near Salt Lake City in winter. (Stansbury.)

*59. *Passerculus savanna* (Wils.), var. *alaudinus*, Bon.—Lark-Sparrow.
Abundant in marshy localities.

*60. *Poocaëtes gramineus* (Gm.), var. *confinis* Bd.—Bay-winged Sparrow.
Very abundant, frequenting the plains.

*61. *Coturniculus passerinus* (Wils.), var. *perpallidus*, Ridg.—Yellow-winged Sparrow.
Rare.

*62. *Chondestes grammaca* (Say).—Lark-Bunting.
Abundant everywhere on plains and benches.

*63. *Zonotrichia leucophrys* (Forst.)—White-crowned Sparrow.
Breeds abundantly in the mountains.

64. *Zonotrichia leucophrys* (Forst.), var. *intermedia*, Ridgw.
Exceedingly abundant in fall, and also a winter-resident.

65. *Junco hyemalis* (L.)—Black Snowbird.
Rare in fall. One specimen only taken in flock of *Zonotrichia intermedia*.

66. *Junco oregonus* (Towns.)—Oregon Snowbird.
Common in fall. Winters at least in southern part of Territory.

*67. *Junco caniceps* (Woodh.)—Red-backed Snowbird.
Tolerably common in the pines of Wahsatch Mountains in the breeding-season. (Ridgway.)

*68, *Poospiza bilineata* (Cass.)—Black-throated Sparrow.
Breeds abundantly in the vicinity of Salt Lake City. (Ridgway.) .

*69. *Poospiza belli* (Cass.), var. *nevadensis*, Ridgw.—Sage Sparrow.
Very common, especially as a winter-resident, frequenting the sagebrush plains.

*70. *Spizella socialis* (Wils.), var. *arizonæ*, Cs.—Chipping-Sparrow.
Not common. "Breeds near Salt Lake City." (Ridgway.)

*71. *Spizella pallida* (Sw.), var. *breweri*, Cass.—Brewer's Sparrow.
Abundant. Permanent resident. Frequents the sage-brush of the benches.

*72. *Melospiza melodia* (Wils.), var. *fallax*, Bd.—Western Song-Sparrow.
Abundant. Permanent resident.

*73. *Melospiza lincolni* (Aud.)—Lincoln's Finch.
Rather uncommon. Found breeding in Parley's Park by Mr. Ridgway.

74. *Melospiza palustris* (Wils.)—Swamp-Sparrow.
Very rare. A single specimen taken in extreme southern part of Utah October 23.

*75. *Passerella townsendi* (Aud.), var. *schistacea*, Bd.—Slate-colored· Sparrow.
Abundant in the mountains. Breeds. (Ridgway.)

76. *Calamospiza bicolor* (Towns.)—White-winged Blackbird.
A single specimen obtained in Parley's Park in July by Mr. Ridgway. A few seen by Dr. Yarrow in Snake Valley on the borders of Utah.

*77. *Hedymeles melanocephalus* (Sw.)—Black-headed Grossbeak.
Very common.

*78. *Cyanospiza amœna* (Say).—Lazuli Finch.
Numerous in the valleys.

*79. *Pipilo maculatus*, Sw., var. *megalonyx*, Bd.—Long-spurred Towhee
Common in the valleys and in chaparral of foot-hills.

*80. *Pipilo chlorurus* (Towns.)—Green-tailed Bunting.
Common. Confined exclusively to the mountains.

81. *Pipilo aberti*, Bd.—Abert's Towhee.
Not rare in extreme southern portion of Utah. Probably breeds.

ALAUDIDÆ (the True Larks.)

*82. *Eremophila alpestris* (Forst.)—Shore-Lark.
Abundant. Permanent resident. Var. *chrysolæma* (Wagl.) Breeding and found sparingly in winter. Var. *leucolæma*, Cs. Predominating in winter. (Ridgway.)

ICTERIDÆ (the Orioles and Blackbirds).

*83. *Molothrus pecoris* (Gm.)—Cow-Bunting.
Not very common.

*84. *Dolichonyx oryzivorus* (L.), var. *albinucha*, Ridg.—Bobolink.
Rather common through the meadows.

*85. *Xanthocephalus icterocephalus* (Bonap.)—Yellow-headed Blackbird
Very numerous. Breeding in large companies. Winters in small numbers.

*86. *Agelæus phœniceus* (L.)—Red-winged Blackbird.
Common resident.

* 87. *Sturnella magna* (L.), var. *neglecta*, Aud.—Western Meadow-Lark.
Very abundant. Permanent resident.

*88. *Icterus bullockii* (Swains.)—Bullock's Oriole.
Abundant. Frequenting the vicinity of the settlements.

*89. *Scolecophagus cyanocephalus* (Wagl.)—Brewer's Blackbird.
· Most abundant of the blackbirds. Permanent resident.

CORVIDÆ (the Crows and Jays).

*90. *Corvus corax*, L., var. *carnivorus*, Bartr.—Raven.
Very abundant. Permanent resident.

*91. *Corvus americanus*, Aud.—Common Crow.
Apparently not common. Seen in vicinity of Provo in July. Of
recent occurrence.

*92. *Picicorvus columbianus* (Wils.)—Nutcracker.
Very common in fall. Inhabits exclusively the mountains. Permanent resident.

*93. *Gymnokitta cyanocephala* (Pr. Max.)—Maximilian's Jay.
Abundant in the cedars. Permanent resident.

*94. *Pica melanoleuca*, Vieill., var. *hudsonica*, Sabine.—Magpie.
Numerous and generally distributed. Resident.

*95. *Cyanura stelleri* (Gmel.), var. *macrolopha*, Bd.—Long-crested Jay.
Common. Found only in the mountains. Resident.

*96. *Cyanocitta floridana* (Bartr.), var. *woodhousii*, Bd.—Woodhouse's
Jay.
Numerous. Resident. Not found in the mountains.

97. *Perisoreus canadensis* (L.), var. *capitalis*, Bd.—Gray Jay.
Wahsatch Mountains. (Allen.)

TYRANNIDÆ (the Tyrant Flycatchers).

*98. *Tyrannus carolinensis* (L.)—The Kingbird.
Quite common near the settlements.

*99. *Tyrannus verticalis*, Say.—Arkansas Flycatcher.
Common.

*100. *Myiarchus crinitus* (L.), var. *cinerascens*, Lawr.
Rare in Parley's Park. (Ridgway.)

*101. *Sayornis sayus* (Bon.)—Say's Flycatcher.
Rather common. Found in the valleys and rocky cañons. (Ridgway.)

*102. *Contopus borealis* (Swains.)—Olive-sided Flycatcher.
Rare in the mountains.

*103. *Contopus virens* (L.), var. *richardsoni*, Swains.—Short-legged
Pewee.

**\*104. *Empidonax flavirentris* Bd., var. *difficilis*, Bd.**—Western Yellow-bellied Flycatcher.
Rare in pine-woods of the mountains in July. (Ridgway.)

**\*105. *Empidonax obscurus* (Swains.)**—Wright's Flycatcher.
Common. Chiefly confined to the mountains.

**\*106. *Empidonax hammondii* (Vesey.)**—Hammond's Flycatcher.
Less common than the preceding. Occurring in the fall.

**\*107. *Empidonax pusillus* (Swains).**—Little Flycatcher.
Especially abundant in the valleys, frequenting the willow-thickets along the streams. "Breeds on the mountains up to 7,000 feet." (Ridgway.)

ALCEDINIDÆ (the Kingfishers).

**\*108. *Ceryle alcyon* (L.).**—Kingfisher.
Common on all the streams. Found by Mr. Ridgway in the mountains up to 7,000 feet.

CAPRIMULGIDÆ (the Goatsuckers).

**\*109. *Chordeiles popetue* (Vieill.), var. *henryi*, Cass.**—Western Nighthawk.
Very abundant in the valleys, and breeding in mountains up to 7,000 feet.

**\*110. *Antrostomus nuttalli* (Aud.)**—Nuttall's Whippoorwill.
Same range as preceding, though much less numerous.

CYPSELIDÆ (the Swifts).

**\*111. *Panyptila saxatilis* (Woodh.)**—White-throated Swift.
Not uncommon. Builds its nest in holes in limestone-cliffs.

TROCHILIDÆ (the Humming-Birds).\*

**\*112. *Trochilus alexandri*, Bourc. and Muls.**—Alexander's Humming-Bird.
Numerous in the valleys. "Breeds up to 8,000 feet." (Ridgway.)

**\*113. *Selasphorus platycercus* (Swains.)**—Broad-tailed Humming-Bird.
Common at Ogden in September. (Allen.) Exceedingly abundant in Wahsatch Mountains from May to August. (Ridgway.)

CUCULIDÆ (the Cuckoos).

**114. *Coccyzus americanus* (L.)**—Yellow-billed Cuckoo.
Heard in July at Provo. As the species breeds abundantly in Arizona (Tucson, Bendire), as well as in Nevada and Sacramento Valley (Ridgway), it doubtless nests in portions of Utah also. The season at which it was noted renders this supposition most probable.

---

\* *Stellula calliope* (Gould). The Star-throated Hummer doubtless occurs in the mountains of Utah. since it was observed plentifully by Mr. Ridgway in the East Humboldt Mountains, in the eastern portion of Nevada, in August and September.

PICIDÆ (the Woodpeckers).

*115. *Picus villosus* L., var. *harrisii* Aud.—Harris's Woodpecker.
Common. Confined generally to the mountains. Permanent resident.

*116. *Picus pubescens* L., var. *gairdneri*, Aud.—Gairdner's Woodpecker.
Rare. Our specimen noted at Provo in November. A few individuals seen by Mr. Ridgway in Wahsatch Mountains in July.

*117. *Sphyrapicus thyroideus* (Cass.)—Brown-headed Woodpecker; Black-breasted Woodpecker; Williamson's Woodpecker.
Rare in the pine-region (Ridgway.)

118. *Melanerpes erythrocephalus* (L.)—Red-headed Woodpecker.
A single individual observed at Salt Lake City in June. (Ridgway.)

*119. *Melanerpes torquatus* (Wils.)—Lewis's Woodpecker.
Not very common, but generally distributed. Resident.

*120. *Colaptes auratus* (L.), var. *mexicanus*, Swains.—Red-shafted Flicker.
Very common everywhere. Resident.

STRIGIDÆ (the Owls).

*121. *Speotyto cunicularia* (Mol.), var. *hypogæa*, Bon.—Prairie-Owl.
Not very common. Resident.

*122. *Bubo virginianus* (Gmel.), var. *arcticus*, Swains.—Great Horned Owl.
Common in the wooded portions. Resident.

*123. *Otus vulgaris* (L.), var. *wilsonianus*, Less.—Long-eared Owl.
Exceedingly abundant in the thick brush along the streams. Resident.

FALCONIDÆ (the Hawks, Eagles, &c.)

*124. *Falco communis*, Gmel., var. *anatum*, Bon.—Duck-Hawk.
Rather common. Resident.

*125. *Falco saker* Schl., var. *polyagrus*, Cass.—Prairie-Falcon.
Somewhat common on the plains. Resident. (Ridgway.)

*126. *Æsalon columbarius* (L.)—Pigeon-Hawk.
Rather frequent. Generally distributed. Resident.

* 127. *Tinnunculus sparverius* (L.)—Sparrow-Hawk.
Very common everywhere. Resident.

*128. *Pandion haliaëtus* (L.), var. *carolinensis*, Gmel.—Fish-Hawk.
Rather rare. Resident.

*129. *Haliaëtus leucocephalus* (L.)—White-headed Eagle.
Rather common. Resident.

*130. *Aquila chrysaëtos* (L.), var. *canadensis*, L.—Golden Eagle.
Rather common in the mountains. Resident.

*131. *Archibuteo lagopus* (Brünn.), var. *sancti-johannis* Penn.—Black Hawk.
Exceedingly abundant in the vicinity of Provo Lake in winter.

*132. *Archibuteo ferrugineus* (Licht.)—California Squirrel-Hawk.
The eggs of this species, together with the parent birds, collected in the vicinity of Ogden, are in the Smithsonian collection.

*133. *Buteo borealis* (Gmel.), var. *calurus*, Cass.
Common. Resident.

*134. *Buteo swainsoni*, Bon.—Swainson's Buzzard.
Very abundant in the mountains. (Ridgway.)

*135. *Nisus cooperi*, (Bon.)—Cooper's Hawk.
Rare. Generally distributed, but chiefly seen in the mountains. Resident. (Ridgway.)

*136. *Nisus fuscus* (Gm.)—Sharp-shinned Hawk.
Common. Resident.

*137. *Circus cyaneus* (L.), var. *hudsonius*, L.—Marsh-Hawk.
Exceedingly abundant in the lowlands. Resident.

CATHARTIDÆ (the American Vultures).

(?)138. *Pseudogryphus californianus* (Cuv.)—Californian Vulture.
Very rare. Two individuals seen near Beaver November 25.

*139. *Rhinogryphus aura* (L.)—Red-headed Vulture; Turkey-Buzzard.
Common. Resident.

COLUMBIDÆ (the Doves, or Pigeons).

*140. *Zenaidura carolinensis* (L.)—Carolina Dove.
Abundant in the valleys. Breeds up to 8,000 feet. (Ridgway.)

TETRAONIDÆ (the Grouse).

*141. *Canace obscura* (Say).—Dusky Grouse.
Abundant. Resident. Confined exclusively to the mountains.

142. *Centrocercus urophasianus* (Bon.)—Sage-Hen.
Very abundant, principally upon the plains, but found in the valleys of the mountains up to 7,000 feet.

*143. *Pediocœtes phasianellus* (L.), var. *columbianus*, Ord.—Sharp-tailed Grouse.

A single company seen about the middle of September in grassy foot-hills near Meadow Creek. (Yarrow.) Resident.

*144. *Bonasa umbellus* (L.), var. *umbelloides*, Douglas.—Ruffed Grouse.

Occurs sparingly in the mountains near Ogden. (Allen.)   Also near Salt Lake City. (Ridgway.) Resident.

PERDICIDÆ (the Quails or Partridges).

*145. *Ortyx virginianus* (L.)—Quail; Bob White.

Introduced near Ogden and Provo. (Allen.)

*146. *Lophortyx californicus* (Shaw).—Californian Quail.

Introduced near Ogden. (Allen.)

*147. *Lophortyx gambeli* (Nutt.)—Gambel's Quail.

Very abundant in southern part of Territory. Resident.

CHARADRIIDÆ (the Plovers).

*148. *Ægialitis vociferus* (L.)—Kildeer-Plover.

Very numerous. Resident.

*149. *Ægialitis cantianus* (Lath.), var. *nivosus*, Cass.—Snowy Plover.

Very abundant on shores of Salt Lake in May. (Ridgway.)

SCOLOPACIDÆ (the Snipes, Sandpipers, &c.)

150. *Gallinago gallinaria* (Gm.), var. *wilsoni*, Temm.—English Snipe.

Abundant.  Found in Parley's Park during the entire summer.  Probably breeds. (Ridgway.)

151. *Macrorhamphus griseus* (Gm.)—Red-breasted Snipe.

Abundant during the fall.  Probably breeds, as it was obtained at Provo in July in full summer-dress.

152. *Tringa alpina* (L.), var. *americana*, Cass.—Red-backed Sandpiper.

Common at Ogden in September. (Allen.)

153. *Actodromas minutilla* (Vieill.)—Least Sandpiper.

A few seen about July 26 at Provo.  Not common at Ogden. (Allen.)

154. *Ereunetes pusillus* (L.)—Semipalmated Sandpiper.

Abundant during the fall-migrations.

155. *Symphemia semipalmata* (Gmel.)

Numerous on south shore of Salt Lake.  Breeding. (Ridgway.)

156. *Totanus melanoleucus* (Gmel.)—Greater Yellowlegs.

Abundant during the fall-migration.

157. *Totanus flavipes* (Gmel.)—Summer Yellowlegs.
Not common. Ogden, September. (Allen.)

158. *Totanus solitarius* (Bp.)
Not common. Ogden, September. (Allen.)

*159. *Tringoides macularius* (L.)—Spotted Sandpiper.
Common along the streams and lakes.

160. *Actiturus bartramius* (Wils.)—Bartram's Field-Plover.
Rather common on Kamas prairies in July. (Ridgway.)

*161. *Numenius longirostris*, Wils.—Long-billed Curlew.
Breeding abundantly on shores and islands of Salt Lake in May and June. (Ridgway.) Abundant during the fall-migration.

RALLIDÆ (the Rails, Gallinules, and Coots).

162. *Rallus elegans*, Aud.—King-Rail.
Said to be uncommon. Ogden. (Allen.)

*163. *Rallus virginianus*, L.—Virginia Rail.
Common in the marshes. Resident.

*164. *Porzana carolina* (L.)—Carolina Rail.
Not so common as preceding. Winters (?)

*165. *Porzana jamaicensis* (Gm.)—Little Black Rail.
Occasional in summer. Parley's Park, June, July, and August. (Ridgway.)

166. *Fulica americana*, Gm.—Coot.
Very abundant. Resident.

PHALAROPODIDÆ (the Phalaropes).

*167. *Steganopus wilsoni* (Sab.)—Wilson's Phalarope.
Common at Salt Lake.

RECURVIROSTRIDÆ (the Avocets and Stilts).

*168. *Recurvirostra americana*, Gm.—American Avocet.
Abundant. Breeding at Salt Lake in June. (Ridgway.)

*169. *Himantopus nigricollis*, Vieill.—Black-necked Stilt.
Same as preceding.

GRUIDÆ (the Cranes).

*170. *Grus canadensis* (L.)—Sandhill-Crane.
Not uncommon.

. TANTALIDÆ (the Ibises).

171. *Tantalus loculator*, L.—Wood-Ibis.
Rather common visitant.

*172. *Ibis guarauna* (L.)—Glossy Ibis.
Common.

*173. *Ibis alba* (L.)—White Ibis.
A few seen at Ogden, September.  (Allen.)  Probably breeds in considerable numbers.

ARDEIDÆ (the Herons).

*174. *Ardea herodias* L.—Great Blue Heron.
Common.  Resident.

175. *Herodias alba* (L.), var. *egretta*, Gmel.—White Heron.
Not uncommon in the fall.

*176. *Botaurus minor* (Gm.)—Bittern.
Common in all parts of the Territory.  Resident.

*177. *Nyctiardea grisea* (L.), var. *nævia*, Bodd.—Night-Heron.
Very common.  Resident.

ANATIDÆ (the Swans, Geese, and Ducks).

178. *Cygnus americanus*, Sharpl.—Whistling Swan.
Jordan River, March.  (Stansbury.)

179. *Anser hyperboreus*, Pal.—Snow-Goose.*
Common winter-resident.

*180. *Branta canadensis* (L.)—Canada Goose.
Immense flocks pass through the Territory in fall, and large numbers winter.

*181. *Anas boschas*, L.—Mallard.
One of the most common ducks.  Breeding abundantly, and wintering in large numbers.

182. *Anas obscurus*, Gm.—Black Duck.
A few seen at Rush Lake in November.  (Yarrow.)

183. *Dafila acuta* (L.)—Pin-tail.
Common in fall.

*184. *Nettion carolinensis* (Gm.)—Green-winged Teal.
Very abundant.

185. *Querquedula discors* (L.)—Blue-winged Teal.
Not nearly as abundant as preceding.  Perhaps breeds.

*186. *Querquedula cyanoptera* (Vieill.)—Red-breasted Teal.
Common summer-resident.  Breeding abundantly in the marshes.

187. *Spatula clypeata* (L.)—Shoveler.
Very common in the fall.

*188. *Chaulelasmus streperus* (L )—Gadwall.
Very abundant. But few winter.

*189. *Mareca americana* (Gm.)—American Widgeon.
Abundant.

190. *Aix sponsa* (L.)—Summer-Duck.
Common. Ogden, September. (Allen.)

191. *Fulix marila* (L.)—Big Blackhead.
Common in fall.

192. *Fulix affinis* (Eyton).—Little Blackhead.
Autumn-migrant. Utah Lake. (Capt. J. H. Simpson.)

*193. *Fulix collaris* (Donovan).—Ring-necked Duck.
Common.

194. *Aythya ferina* (L.), var. *americana*, Eyton.
Numerous in fall.

195. *Bucephala clangula* (L.), var. *americana*, Bon.—Golden-eye.
Abundant in fall and winter.

196. *Bucephala islandica* Gm.—Barrow's Golden-eye.
Perhaps not uncommon in fall and winter. A pair were taken in Provo River December 1.

197. *Bucephala albeola* (L.)—Butter-ball.
Very common in fall and winter.

*198. *Erismatura rubida* (Wils.)—Ruddy-Duck.
Common.

199. *Mergus merganser*, L., var. *americanus*, Cass.—Sheldrake.

*200. *Mergus serrator*, L.—Red-breasted Merganser.
Abundant.

201. *Lophodytes cucullatus* (L.)—Hooded Merganser.
Common in fall.

### PELECANIDÆ (the Pelicans).

*202. *Pelécanus erythrorhynchus* (Gm.)—American Pelican.
Common upon the lakes. Although no longer breeding upon Great Salt Lake, it undoubtedly does so within the limits of the Territory.

### PHALACROCORACIDÆ (the Cormorants).

*203. *Graculus dilophus* (Sw.)—Double-crested Cormorant; Black Shag.
Common at Salt Lake.

LARIDÆ (the Gulls and Terns).

*204. *Larus argentatus*, Brunn., var. *californicus*, Lawr.—California Herring-Gull.
Common summer-resident. (Ridgway.)

205. *Larus delawarensis*, Ord.—Ring-billed Gull.
Rather common. Winter-resident.

206. *Chrœcocephalus philadelphia* (Ord).
Ogden, October 1. (Allen.).

207. *Xema sabinei* (Sab.)—Fork-tailed Gull.
One taken at Ogden September 28. (Allen.)

*208. *Thalasseus regius*, (Gamb.)—Royal Tern.
Not uncommon in summer. (Ridgway.)

*209. *Sterna forsteri*, Nutt.—Forster's Tern.
Abundant. "Breeds in marshes of Salt Lake." (Ridgway.)

*210. *Hydrochelidon fissipes* (L.)—Short-tailed Tern.
Rather uncommon. "Breeds in marshes of Salt Lake." (Ridgway.)

COLYMBIDÆ (the Loons).

211. *Colymbus glacialis*, L., var. *torqatus*, Brünn.—Great Northern Diver.
Probably not of infrequent occurrence.

PODICIPIDÆ (the Grebes).

212. *Podiceps occidentalis* (Lawr.)—Western Grebe.
Common. Probably breeds.

213. *Podiceps cornutus* (Gm.)—Horned Grebe.
Rather common in fall.

214. *Podilymbus podiceps* (L.)—Carolina Grebe.
Common in fall.

# REPORT UPON AND LIST OF BIRDS COLLECTED BY THE EXPEDITION FOR GEOGRAPHICAL AND GEOLOGICAL EXPLORATIONS AND SURVEYS WEST OF THE ONE HUNDREDTH MERIDIAN IN 1873, LIEUT. G. M. WHEELER, CORPS OF ENGINEERS, IN CHARGE.

By H. W. HENSHAW, APRIL, 1874.

The report presented in the following pages is based upon the material gathered during the field-season of 1873, in connection with the geographical and geological survey west of the one hundredth meridian, made under the auspices of the Engineer Department, Lieut. G. M. Wheeler commanding. It includes not only my own work, but also the results in this department of both Drs. Rothrock and Newberry, jr., by whose joint labors some two hundred birds were added to the collection, and to each of whom I am indebted for certain information respecting the habits and range of species noted by them, some of which were not met with by myself. All such information has been made available, and will be found accredited to its proper source. The season's collection of birds amounted to very nearly twelve hundred specimens, representing over two hundred species. Others also were noted, and their identity ascertained beyond doubt, of which no specimens were secured. For convenience of reference, the report is divided into three sections; this course being rendered necessary by the wide separation of the localities at which the larger portion of the work was done. The first contains the observations made at Denver from May 5 to 22; the second, those made at Fort Garland, Southern Colorado, including also much of the information gathered by Dr. Rothrock in the mountains of Colorado; while in the third portion are given the results of the joint labors of both Dr. Newberry, jr., and myself, extending over a very large area of country, principally in Eastern Arizona and Western New Mexico, and covering an interval of time from July 15 till November 25. By the kind permission of Lieutenant Wheeler I was enabled to make an early start, and arrived in Denver the 5th of May, intending to proceed directly to Fort Garland, a locality which had been selected as affording a promising field for natural-history work, more especially in ornithology. Through the unavoidable detention of my collecting-material, my stay in Denver was prolonged for more than two weeks. This interval till the 22d of May was spent in making daily excursions in the vicinity of the city, more especially along the banks of the Platte River, which is here tolerably well timbered, principally with cottonwoods, and on Cherry Creek. At the time of my arrival I found the season quite backward, and the vegetation was little, if any, in advance of what I left in the vicinity of Boston. But few of the trees had fairly begun to leaf out, though before my departure the cottonwoods and many others were far advanced in this respect. The observations made at this time are believed to be possessed of very considerable value as giving the time of arrival of quite a large number of species, while the capture of quite a number is of especial interest, as extending their range much farther to the west than was hitherto known. The fauna in the vicinity of Denver is perhaps best compared with the Carolinian of the Eastern Province; but the list presents quite a number, as would naturally be expected from the early season at which the collection was made, which are to be regarded merely as migrants, and which spend the summer ar to the northward. The arrangement and nomenclature is in most cases that adopted in the "Birds of North America," by Baird, Brewer, and Ridgway.

## SECTION I.

### TURDIDÆ (the Thrushes).

1. *Turdus migratorius*, L.

Quite common; a pair seen May 10, building their nest in the partially open cavity of a tree.

2. *Turdus fuscescens*, Stephens.—Tawny Thrush.

First seen May 17, after which it was daily noticed in small numbers frequenting the moist thickets bordering the small creeks.

| No. | Sex. | Date. | Collector. | Wing. | Tail. | Bill. | Tarsns. |
|---|---|---|---|---|---|---|---|
| 108 | ♂ ad. | May 17, 1873 | Henshaw .......... | 4. 18 | 3. 28 | 0. 55 | 1. 12 |

3. *Turdus swainsoni*, Cabanis.—Olive-backed Thrush.

Several noticed May 12. By the 17th, this species fairly swarmed in the same localities as the preceding. The females were apparently a full week later than the males.

| No. | Sex. | Date. | Collector. | Wing. | Tail. | Bill. | Tarsus. |
|---|---|---|---|---|---|---|---|
| 49 | ♂ ad. | May 12, 1873 | Henshaw ........... | 3. 93 | 2. 95 | 0. 51 | 1. 09 |
| 109 | ♂ ad. | May 17, 1873 | .... do ........... | 3. 90 | 3. 02 | 0. 55 | 1. 04 |
| 110 | ♂ ad. | May 17, 1873 | .... do ........... | 3. 98 | 3. 07 | 0. 49 | 1. 09 |
| 111 | ♂ ad. | May 17, 1873 | .... do ........... | 4. 04 | 3. 05 | 0. 52 | 1. 02 |
| 112 | ♂ ad. | May 17, 1873 | .... do ........... | 4. 00 | 2. 93 | 0. 48 | 1. 09 |
| 113 | ♂ ad. | May 17, 1873 | .... do ........... | 4. 10 | 3. 11 | 0. 50 | 1. 03 |
| 114 | ♂ ad. | May 17, 1873 | .... do ........... | 3. 30 | 2. 93 | 0. 50 | 1. 00 |
| 115 | ♂ ad. | May 17, 1873 | .... do ........... | 4. 02 | 2. 98 | 0. 48 | 1. 08 |
| 118 | ♂ ad. | May 17, 1873 | .... do ........... | 3. 98 | 2. 93 | 0. 50 | 1. 12 |

4. *Galeoscoptes carolinensis*, (L.)—Catbird.

Apparently not common; one seen May 14.

| No. | Sex. | Date. | Collector. | Wing. | Tail. | Bill. | Tarsus. . |
|---|---|---|---|---|---|---|---|
| 88 | ♂ ad. | May 14, 1873 | Henshaw .......... | 3. 75 | 4. 20 | 0. 63 | 1. 03 |

5. *Harporhynchus rufus*, (L.)—Brown Thrasher.
A single individual seen May 10. Noted here, also, by Dr. Rothrock.

| No. | Sex. | Date. | Collector. | Wing. | Tail. | Bill. | Tarsus. |
|---|---|---|---|---|---|---|---|
| 13 | ........ | ............... | Rothrock ........... | 4.04 | 5.07 | 0.85 | 1.31 |
| 60 | ♂ ad. | May 12, 1873 | Henshaw .......... | 4.42 | 5.71 | 1.02 | 1.30 |

SYLVIIDÆ (the Sylvias).

6. *Regulus calendula*, (L.)
Present in small numbers, usually one or more accompanying each flock of Chickadees.

PARIDÆ (the Titmice).

7. *Parus atricapillus*, L., var. *septentrionalis*, Harris.—Long-tailed Chickadee.
Quite frequently seen in small flocks.

TROGLODYTIDÆ (the Wrens).

8. *Troglodytes aëdon*, V., var. *parkmanni*, Aud.—Parkman's Wren.
One seen May 6, and afterward an occasional individual noticed. More frequently heard than seen, as it frequents the thickest clump of bushes and patches of briers, where a glimpse may now and then be had of it as it glides along.

| No. | Sex. | Date. | Collector. | Wing. | Tail. | Bill. | Tarsus. |
|---|---|---|---|---|---|---|---|
| 82 | ♂ ad. | May 14, 1873 | Henshaw .......... | 2.09 | 1.97 | 0.50 | 0.68 |
| 132 | ♂ ad. | May 22, 1873 | .... do .......... | 2.10 | 2.02 | 0.54 | 0.63 |

SYLVICOLIDÆ (the Warblers).

9. *Helminthophaga celata*, (Say.)—Orange-crowned Warbler.
After May 7, when this species was first seen, it was not very common. It is an active insect-hunter, and frequents alike the tops of the smaller trees and the low bushes, where it may be often seen darting forth in pursuit of some passing insect.

| No. | Sex. | Date. | Collector. | Wing. | Tail. | Bill. | Tarsus. |
|---|---|---|---|---|---|---|---|
| 24 | ♂ ad. | May 9, 1873 | Henshaw .......... | 2.40 | 2.09 | 0.29 | 0.66 |
| 94 | ♂ ad. | May 17, 1873 | ....do ............. | 2.47 | 2.14 | 0.40 | 0.64 |

10. *Dendroica æstiva*, (Gm.)—Yellow Warbler.

Common everywhere. Not seen till May 14, when it made its appearance in large numbers.

| No. | Sex. | Date. | Collector. | Wing. | Tail. | Bill. | Tarsus. |
|---|---|---|---|---|---|---|---|
| 92 | ♀ ad. | May 15, 1873 | Henshaw .......... | 2.37 | 1.95 | 0.40 | 0.70 |

11. *Dendroica audubonii*, (Towns.)—Audubon's Warbler.

Small numbers of this and the following species were seen May 7 ; common on the 10th. Its habits and notes appear to correspond almost exactly with those of the common Yellow Rump.

| No. | Sex. | Date. | Collector. | Wing. | Tail. | Bill. | Tarsus. |
|---|---|---|---|---|---|---|---|
| 9 | ♂ ad. | May 7, 1873 | Henshaw ....... ... | 3.17 | 2.50 | 0.43 | 0.76 |
| 10 | ♂ ad. | May 7, 1873 | ....do ............. | 3.08 | 2.50 | 0.47 | 0.75 |
| 22 | ♂ ad. | May 9, 1873 | ....do ............. | 3.14 | 2.54 | 0.45 | 0.75 |
| 37 | ♂ ad. | May 10, 1873 | ....do ............. | 3.00 | 2.42 | 0.43 | 0.72 |
| 36 | ♂ ad. | May 11, 1873 | ....do ............. | 3.03 | 2.45 | 0.42 | 0.72 |
| 38 | ♂ ad. | May 11, 1873 | ....do ............. | 3.07 | 2.47 | 0.45 | 0.70 |
| 39 | ♀ ad. | May 11, 1873 | ....do ............. | 2.90 | 2.38 | 0.40 | 0.74 |
| 68 | ♂ ad. | May 13, 1873 | ....do ............. | 3.16 | 2.61 | 0.45 | 0.74 |
| 120 | ♀ ad. | May 17, 1873 | ....do ............. | 3.11 | 2.51 | 0.45 | 0.71 |

12. *Dendroica coronata*, (L.)—Yellow-rump Warbler.

In much fewer numbers than the preceding, with which it was asso-ciated. Have heard males of the two species singing in the same tree. All apparently migrate farther north.

| No. | Sex. | Date. | Collector. | Wing. | Tail. | Bill. | Tarsus. |
|---|---|---|---|---|---|---|---|
| 11 | ♀ ad. | May 7, 1873 | Henshaw .......... | 2.80 | 2.33 | 0.40 | 0.72 |
| 23 | ♂ ad. | May 9, 1873 | ....do ............. | 2.90 | 2.35 | 0.39 | 0.73 |
| 123 | ♀ ad. | May 17, 1873 | ....do ............. | 2.90 | 2.33 | 0.40 | 0.73 |

13. *Dendroica maculosa*, (Gm.)—Black and Yellow Warbler.

A single fine male taken May 17. No others were seen. This is the first note of its occurrence west of the plains.

| No. | Sex. | Date. | Collector. | Wing. | Tail. | Bill. | Tarsus. |
|---|---|---|---|---|---|---|---|
| 93 | ♂ ad. | May 17, 1873 | Henshaw .......... | 2.40 | 2.15 | 0.45 | 0.66 |

14. *Dendroica cærulea*, (Wils.)—Cœrulean Warbler.

A small warbler seen May 17 was unquestionably of this species. Its small size and bright-blue color made it conspicuous among a flock of Audubon's warblers, as they passed rapidly from tree to tree, but my attention being diverted for a moment I lost sight of it, nor was it again seen. Not hitherto detected west of the plains. "Apparently common at Leavenworth, Kansas" (Allen).

15. *Dendroica striata*, (Forst.)—Black-poll Warbler.

Both sexes abundant May 17. This is, I believe, the most western locality at which the species has been recorded.

| No. | Sex. | Date. | Collector. | Wing. | Tail. | Bill. | Tarsus. |
|-----|------|-------|------------|-------|-------|-------|---------|
| 120 | ♀ ad. | May 17, 1873 | Henshaw .......... | 2.83 | 2.22 | 0.42 | 0.73 |

16. *Seiurus noveboracensis*, (Gm.)—Small-billed Water-Thrush.

One specimen secured May 12. Afterward observed in small numbers, frequenting the margins of pools and streams as at the East.

| No. | Sex. | Date. | Collector. | Wing. | Tail. | Bill. | Tarsus. |
|-----|------|-------|------------|-------|-------|-------|---------|
| 53 | ♀ ad. | May 12, 1873 | Henshaw .......... | 3.04 | 2.25 | 0.55 | 0.81 |

17. *Geothlypis trichas*, (L.)—Maryland Yellowthroat.

Apparently not common. A female shot May 7.

| No. | Sex. | Date. | Collector. | Wing. | Tail. | Bill. | Tarsus. |
|-----|------|-------|------------|-------|-------|-------|---------|
| 12 | ♀ ad. | May 7, 1873 | Henshaw .......... | 2.12 | 2.13 | 0.45 | 0.71 |

18. *Geothlypis macgillivrayi*, (Aud.)—Macgillivray's Warbler.

A single male taken May 14. Both sexes common a few days later. Brush-heaps form a favorite hunting-ground for this species.

| No. | Sex. | Date. | Collector. | Wing. | Tail. | Bill. | Tarsus. |
|-----|------|-------|------------|-------|-------|-------|---------|
| 80 | ♂ ad. | May 14, 1873 | Henshaw .......... | 2.47 | 2.34 | 0.47 | 0.77 |
| 117 | ♂ ad. | May 17, 1873 | ....do .............. | 2.43 | 2.30 | 0.46 | 0.82 |
| 122 | ♂ ad. | May 18, 1873 | ....do .............. | 2.48 | 2.38 | 0.45 | 0.82 |

19. *Myiodioctes pusillus*, (Wils.)—Green Black-capped Flycatcher.

Seen May 14. Common among the shrubbery and trees that skirt the small streams and ponds.

| No. | Sex. | Date. | Collector. | Wing. | Tail. | Bill. | Tarsus. |
|-----|------|-------|------------|-------|-------|-------|---------|
| 81 | ♀ ad. | May 14, 1873 | Henshaw .......... | 2.21 | 2.14 | 0.40 | 0.70 |

HIRUNDINIDÆ (the Swallows).

20. *Hirundo horreorum*, Bartou.—Barn-Swallow.
Not very numerous.

21. *Petrochelidon lunifrons*, (Say.)—Cliff-Swallow.
A few pairs only seen.

22. *Tachycineta thalassina*, (Sw).—Violet-green Swallow.
A few noted the 12th of May.

23. *Stelgidopteryx serripennis*, (Aud.)
A few seen along Cherry Creek the 6th of May. Common about the 12th.

LANIIDÆ (the Shrikes).

24. *Collurio ludovicianus*, (L.), var. *excubitoroides*, Sw.—White-rumped Shrike.
Numerous individuals of this species were seen during the first days of May, and apparently all were mated, and possibly nesting, though I did not succeed in finding any nests. It has at this season quite a number and variety of notes, some of which are the call-notes and common to both sexes. The male also makes an occasional attempt at a song, and the notes, though harsh, are not unpleasing.

| No. | Sex. | Date. | Collector. | Wing. | Tail. | Bill. | Tarsus. |
|-----|------|-------|-----------|-------|-------|-------|---------|
| 20 | ♀ ad. | May 7, 1873 | Henshaw .......... | 3.63 | 3.90 | 0.63 | 01.02 |
| 62 | ♂ ad. | May 12, 1873 | ....do .............. | 3.85 | 4.23 | 0.56 | 01.04 |

TANAGRIDÆ (the Tanagers).

25. *Pyranga ludoviciana*, (Wils.)—Louisiana Tanager.
But a single individual seen, May 20.

26. *Pyranga æstiva*, (L.), var. *cooperi*, Ridg.—Cooper's Tanager.
A single male shot May 10.

| No. | Sex. | Date. | Collector. | Wing. | Tail. | Bill. | Tarsus. |
|-----|------|-------|-----------|-------|-------|-------|---------|
| 54 | ♂ young of year. | May 12, 1873 | Henshaw .... .... | 3.61 | 3.00 | 0.80 | 0.77 |

FRINGILLIDÆ (the Finches).

27. *Chrysomitris tristis*, (L.)
Very abundant in large flocks in the cottonwood-groves along the Platte River. These fairly resounded with the twitterings and chirp-

ings of the young males, which appeared to be practicing for the full concerts that follow later. Both sexes were molting and in curiously-pied plumage.

| No. | Sex. | Date. | Collector. | Wing. | Tail. | Bill. | Tarsus. |
|---|---|---|---|---|---|---|---|
| 7 | ♂ | May 6, 1873 | Henshaw ........... | 2.87 | 2.02 | 0.45 | 0.56 |
| 100 | ♂ ad. | May 6, 1873 | ....do .............. | 2.70 | 2.06 | o.47 | 0.52 |

28. *Passerculus savanna,* (Wils.), var. *alaudinus,* Bp.—Savannah Sparrow.

Common the 1st of May, and noted in increasing numbers till the 12th, when they were exceedingly numerous around the small ponds and marshy spots.

| No. | Sex. | Date. | Collector. | Wing. | Tail. | Bill. | Tarsus. |
|---|---|---|---|---|---|---|---|
| 52 | ♂ ad. | May 12, 1873 | Henshaw ........... | 2.83 | 2.23 | 0.43 | 0.78 |
| 121 | ♂ ad. | May 17, 1873 | ....do .............. | 2.80 | 2.15 | 0.43 | 0.80 |

29. *Pocœcetes gramineus,* (Gm.), var. *confinis,* Bd.—Grass-Finch; Bay-winged Bunting.

Common and in full song on my first arrival. The song does not differ from that of its eastern representative, from which the present variety is distinguished by its paler colors and somewhat slenderer bill.

| No. | Sex. | Date. | Collector. | Wing. | Tail. | Bill. | Tarsus. |
|---|---|---|---|---|---|---|---|
| 18 | ♂ ad. | May 7, 1873 | Henshaw ........... | 3.25 | 2.53 | 0.48 | 0.82 |
| 19 | ♀ ad. | May 7, 1873 | ....do .............. | 3.17 | 2.67 | 0.45 | 0.73 |
| 41 | ♂ ad. | May 10, 1873 | ....do .............. | 3.13 | 2.55 | 0.45 | 0.83 |

30. *Chondestes grammaca,* (Say).—Lark-Finch.

Very common in small companies along the banks of the Platte. During the vernal season its beautiful warbling song is hardly excelled by any other species.

| No. | Sex. | Date. | Collector. | Wing. | Tail. | Bill. | Tarsus. |
|---|---|---|---|---|---|---|---|
| 3 | ♂ ad. | May 6, 1873 | Henshaw ........ .... | 3.65 | 3.05 | 0.50 | 0.80 |
| 17 | ♂ ad. | May 7, 1873 | ....do .......... .... | 3.43 | 2.90 | 0.50 | 0.75 |
| 27 | ♂ ad. | May 9, 1873 | ....do .......... .... | 3.48 | 3.08 | 0.50 | 0.78 |
| 12 | ♂ ad. | June —, 1873 | Dr. Rothrock........ | 3.20 | 2.79 | 0.50 | 0.72 |
| 5 | ♂ | June —, 1873 | ....do ............. | 3.45 | 2.88 | 0.52 | 0.77 |

31. *Zonotrichia leucophrys,* (Forst.)—White-crowned Sparrow.

Quite common in small flocks from the 7th till the 20th. Dr. Rothrock found it breeding in the South Park in July.

| No. | Sex. | Date. | Collector. | Wing. | Tail. | Bill. | Tarsus. |
|-----|------|-------|------------|-------|-------|-------|---------|
| 4 | ♂ ad. | May 6, 1873 | Henshaw ........ ...... | 3.02 | 3.16 | 0.43 | 0.92 |
| 16 | ♂ ad. | May 7, 1873 | ....do ........ ........ | 3.15 | 3.32 | 0.43 | 0.86 |
| 42 | ♂ ad. | May 10, 1873 | ....do ........ ........ | 3.16 | 3.12 | 0.44 | 0.93 |
| 87 | ♂ ad. | May 14, 1873 | ....do ........ ........ | 3.12 | 3.12 | 0.43 | 0.85 |
| 107 | ♀ ad. | May 17, 1873 | ....do ........ ........ | 2.95 | 3.00 | 0.43 | 0.85 |
| 67 | Ad. | June 26, 1873 | Rothrock........... | 3.18 | 3.30 | 0.50 | 0.95 |

32. *Zonotrichia leucophrys*, (Forst), var. *intermedia*, Ridgw.—Gambel's Finch.

Under the variety *intermedia* Mr. Ridgway distinguishes the Middle Province form from the true *gambeli* as restricted to the Pacific coast. This variety is to be known by its lighter coloration and the chestnut-brown dorsal streaks instead of black. Up to the 10th of May a few individuals were seen accompanying flocks of the preceding, but they were evidently stragglers, the main body having passed on earlier. None apparently remain to breed, but all pass further north.

| No. | Sex. | Date. | Collector. | Wing. | Tail. | Bill. | Tarsus. |
|-----|------|-------|------------|-------|-------|-------|---------|
| 15 | ♀ ad. | May 7, 1873 | Henshaw ........ .... | 2.96 | 3.08 | 0.46 | 0.82 |

33. *Spizella socialis*, (Wils.)—Chipping-Sparrow.

Rather common in flocks after the 6th.

| No. | Sex. | Date. | Collector. | Wing. | Tail. | Bill. | Tarsus. |
|-----|------|-------|------------|-------|-------|-------|---------|
| 5 | ♂ ad. | May 6, 1873 | Henshaw ........ .... | 2.77 | 2.52 | 0.40 | 0.68 |

34. *Spizella pallida*, (Sw.), var. *breweri*, Cass.—Brewer's Sparrow.

A single specimen, the only one seen, taken the 17th.

| No. | Sex. | Date. | Collector. | Wing. | Tail. | Bill. | Tarsus. |
|-----|------|-------|------------|-------|-------|-------|---------|
| 116 | ♀ ad. | May 17, 1873 | Henshaw ........ .... | 2.32 | 2.52 | 0.38 | 0.63 |

35. *Melospiza melodia*, (Wils.), var. *fallax*.—Western Song-Sparrow.

But few of this species seen. With the exception of its generally paler colors and more slender bill, it presents little or no differences from the eastern form (*melodia*), and in habits and notes the two are absolutely identical.

| No. | Sex. | Date. | Collector. | Wing. | Tail. | Bill. | Tarsus. |
|-----|------|-------|------------|-------|-------|-------|---------|
| 40 | ♀ ad. | May 10, 1873 | Henshaw ........ .... | 2.55 | 2.83 | 0.45 | 0.82 |
| 14 | ♀ ad. | May 7, 1873 | ....do ........ .... | 2.60 | 2.71 | 0.48 | 0.80 |

36. *Melospiza lincolni*, (Aud.)—Lincoln's Finch.

A few seen between the 7th and 11th of May. By the 17th this was one of the commonest birds, outnumbering all the other sparrows. It was found in almost every clump of bushes and grove of trees, but prefers moist thickets.

| No. | Sex. | Date. | Collector. | Wing. | Tail. | Bill. | Tarsus. |
|---|---|---|---|---|---|---|---|
| 13 | ♂ ad. | May 7, 1873 | Henshaw ............. | 2.41 | 2.46 | 0.45 | 0.79 |
| 84 | ♂ ad. | May 14, 1873 | ....do ............... | 2.52 | 2.57 | 0.44 | 0.81 |
| 86 | ♂ ad. | May 14, 1873 | ....do ............... | 2.57 | 2.65 | 0.45 | 0.80 |
| 91 | ♀ ad. | May 15, 1873 | ....do ............... | 2.25 | 2.33 | 0.45 | 0.75 |
| 100 | ♂ ad. | May 17, 1873 | ....do ............... | 2.48 | 2.48 | 0.45 | 0.83 |
| 101 | ♀ ad. | May 17, 1873 | ....do ............... | 2.64 | 2.67 | 0.45 | 0.82 |
| 102 | ♂ ad. | May 17, 1873 | ....do ............... | 2.23 | 2.26 | 0.44 | 0.73 |

37. *Calamospiza bicolor*, (Towns.)—Lark-Bunting.

Small flocks seen about ten miles south of the city.

38. *Hedymeles melanocephalus*, (Sw.)—Black-headed Grossbeak.

A male seen the 14th, and a female the 17th. Undoubtedly a common species later.

39. *Cyanospiza amœna*, (Say.)—Lazuli-Finch.

A single male noted the 17th. Observed on several occasions afterward.

| No. | Sex. | Date. | Collector. | Wing. | Tail. | Bill. | Tarsus. |
|---|---|---|---|---|---|---|---|
| 119 | ♂ ad. | May 17, 1873 | Henshaw .......... .... | 2.98 | 2.42 | 0.38 | 0.65 |

40. *Pipilo erythrophthalmus*, (L.), var. *megalonyx*, Bd.—Long-spurred Towhee.

Very common in the brush in early May. Its call-note is so exactly like that of our common cat-bird as to readily deceive one as to the originator. It song, too, presents decided differences from that of our eastern towhee (*P. erythrophthalmus*).

| No. | Sex. | Date. | Collector. | Wing. | Tail. | Bill. | Tarsus. |
|---|---|---|---|---|---|---|---|
| 1 | ♂ ad. | May 6, 1873 | Henshaw .......... .... | 3.53 | 4.15 | 0.56 | 1.05 |
| 2 | ♂ ad. | May 6, 1873 | ....do ............... | 3.34 | 4.05 | 0.54 | 1.10 |
| 28 | ♂ ad. | May 9, 1873 | ....do ............... | 3.48 | 4.42 | 0.55 | 1.10 |

41. *Pipilo chlorurus*, (Towns.)—Green-tailed Bunting.

Two seen for first time May 10; after which time they were rather common.

| No. | Sex. | Date. | Collector. | Wing. | Tail. | Bill. | Tarsus. |
|---|---|---|---|---|---|---|---|
| 43 | ♂ ad. | May 10, 1873 | Henshaw ............ | 3.07 | 3.66 | 0.50 | 0.93 |
| 44 | ♂ ad. | May 10, 1873 | ....do .............. | 3.13 | 3.55 | 0.54 | 0.96 |
| 69 | ♀ ad. | May 13, 1873 | ....do .............. | 2.98 | 3.38 | 0.50 | 0.95 |
| 70 | ♀ ad. | May 13, 1873 | ....do .............. | 2.95 | 3.40 | 0.53 | 0.87 |
| 106 | ♂ ad. | May 17, 1873 | ....do .............. | 3.14 | 3.60 | 0.53 | 0.95 |
| 150 | ♂ ad. | May 27, 1873 | ....do .............. | 2.95 | 3.35 | 0.50 | 1.00 |

ALAUDIDÆ (the Larks).

42. *Eremophila alpestris*, (Forst.), var. *chrysolæma*, Wagl.

Quite numerous on the plains in the neighborhood of the city. Has a rather feeble, but pleasing, warbling song at this season, which the birds uttered while perched on a fence-rail or from the ground.

| No. | Sex. | Date. | Collector. | Wing. | Tail. | Bill. | Tarsus. |
|---|---|---|---|---|---|---|---|
| 29 | ♂ ad. | May 9, 1873 | Henshaw .......... | 4.12 | 2.95 | 0.50 | 0.80 |
| 30 | ♂ ad. | May 9, 1873 | ....do .............. | 4.14 | 3.03 | 0.52 | 0.82 |

ICTERIDÆ (the Orioles).

43. *Molothrus pecoris*, (Gm.)—Cow-Blackbird.

Seen in small flocks of six or seven during my whole stay in the vicinity.

| No. | Sex. | Date. | Collector. | Wing. | Tail. | Bill. | Tarsus. |
|---|---|---|---|---|---|---|---|
| 8 | ♂ ad. | May 6, 1873 | Henshaw .......... | 4.67 | 3.34 | 0.73 | 1.08 |
| 31 | ♂ ad. | May 9, 1873 | ....do ............ | 4.50 | 3.22 | 0.66 | 1.04 |
| 33 | ♂ ad. | May 9, 1873 | ....do ............ | 4.38 | 3.29 | 0.70 | 1.07 |

44. *Agœlæus phœnicens*, (L.)—Red-winged Blackbird.

Both sexes common May 6. By the 17th a few pairs had selected the sites for their nests, and were about to build. A female taken here has a conspicuous bright-red shoulder-patch, streaked slightly with black and bordered with yellowish-white.

| No. | Sex. | Date. | Collector. | Wing. | Tail. | Bill. | Tarsus. |
|---|---|---|---|---|---|---|---|
| 67 | ♀ ad. | May 12, 1873 | Henshaw ...... .... | 4.21 | 3.45 | 0.73 | 0.97 |
| 7 | ♂ ad. | June —, 1873 | Rothrock ...... .... | 5.03 | 4.20 | 0.92 | 1.12 |

45. *Xanthocephalus icterocephalus*, (Bon.)—Yellow-headed Blackbird.

Observed by Dr. Rothrock as quite common on the edges of swamps and in the cottonwoods, June 3.

| No. | Sex. | Date. | Collector. | Wing. | Tail. | Bill. | Tarsus. |
|---|---|---|---|---|---|---|---|
| 1 | ♀ ad. | June 3, 1873 | Dr. Rothrock....... | 5.29 | 3.71 | 0.85 | 1.35 |

46. *Sturnella magna* (L.), var. *neglecta*, Aud.—Western Meadow-Lark.
Common in the fields. The differences of song between this variety and our eastern lark (*magna*) are very striking; so much so that farmers from the east seem very generally to recognize this dissimilarity.

47. *Icterus bullockii* (Sw.)—Bullock's Oriole.
The males made their appearance about the 10th, and the females a few days later. Very common, usually keeping in the tops of the tallest trees.

| No. | Sex. | Date. | Collector. | Wing. | Tail. | Bill. | Tarsus. |
|---|---|---|---|---|---|---|---|
| 71 | ♂ ad. | May 13, 1873 | Henshaw ............ | 4.21 | 3.45 | 0.75 | 0.95 |
| 72 | ♂ ad. | May 13, 1873 | ....do ............... | 3.91 | 2.27 | 0.74 | 0.97 |
| 73 | ♂ ad. | May 13, 1873 | ....do ............... | 3.98 | 3.31 | 0.75 | 0.94 |
| 74 | ♂ ad. | May 13, 1873 | ....do ............... | 4.25 | 3.61 | 0.74 | 0.95 |
| 75 | ♀ ad. | May 13, 1873 | ....do ............... | 3.87 | 3.82 | 0.80 | 0.92 |
| 152 | ♂ ad. | May 27, 1873 | ....do ............... | 4.08 | 3.63 | 0.77 | 0.87 |
| ·3 | ♂ ad. | June —, 1873 | Dr. Rothrock....... | 4.05 | 3.41 | 0.73 | 0.94 |
| 3a | ♂ ad. | June —, 1873 | ....do ............... | 4.00 | 3.33 | 0.75 | 0.90 |

48. *Scolecophagus cyanocephalus* (Wagl.)—Brewer's Blackbird.
Seen the 6th. Apparently not very common.

49. *Quiscalus purpureus* (Bartr.), var. *æneus*, Ridg.—Bronzed Grakle.
Rather numerous May 14.

| No. | Sex. | Date. | Collector. | Wing. | Tail. | Bill. | Tarsus. |
|---|---|---|---|---|---|---|---|
| 89 | ♂ ad. | May 14, 1873 | Henshaw .......... | 5.60 | 4.85 | 1.28 | 1.43 |

CORVIDÆ (the Crows).

50. *Corvus corax*, L., var. *carnivorus*, Bartr.—American Raven.
Common everywhere. Subsists largely upon carrion, and is always to be found in the neighborhood of the slaughter-houses.

51. *Pica melanoleuca* (V.), var. *hudsonica*, Sab.—Magpie.
Rather common in the timber along the streams. Also eminently carnivorous.

TYRANNIDÆ (the Flycatchers).

52. *Tyrannus carolinensis* (L.)—Kingbird.
Arrived the 7th. Not very common. Seemed to prefer the open plain, where it perched upon the tall weeds, to the wooded districts along the streams, most frequented by the following species.

53. *Tyrannus verticalis*, Say.—Arkansas Flycatcher.
One seen the 7th. Common afterward. Very bold, noisy, and quarrelsome. The males at this season were constantly fighting among

themselves, but on the appearance of a raven or hawk, all would unite and make a combined and always successful attack upon the intruder.

| No. | Sex. | Date. | Collector. | Wing. | Tail. | Bill. | Tarsus. |
|-----|------|-------|------------|-------|-------|-------|---------|
| 58 | ♂ ad. | May 12, 1873 | Henshaw | 5.27 | 4.00 | 0.75 | 0.77 |
| 63 | ♂ ad. | May 12, 1873 | ....do | 5.23 | 4.20 | 0.78 | 0.75 |
| 76 | ♂ ad. | May 12, 1873 | ....do | 5.30 | 4.11 | ........ | 0.77 |
| 77 | ♂ ad. | May 13, 1873 | ....do | 5.20 | 4.07 | 0.80 | 0.75 |
| 78 | ♀ ad. | May 13, 1873 | ....do | 4.83 | 4.90 | 0.70 | 0.70 |
| 79 | ♂ ad. | May 13, 1873 | ....do | 5.07 | 4.88 | 0.76 | 0.75 |
| 4 | ........ | June —, 1873 | Dr. Rothrock | 5.09 | 4.21 | 0.75 | 0.74 |
| 4a | ♂ | June —, 1873 | ....do | 4.98 | 3.90 | 0.73 | 0.68 |

54. *Sayornis sayus* (Bon.)—Say's Flycatcher.
A specimen taken the 9th. Rather uncommon. Habits very similar to those of the eastern Pewee (*fuscus*).

| No. | Sex. | Date. | Collector. | Wing. | Tail. | Bill. | Tarsus. |
|-----|------|-------|------------|-------|-------|-------|---------|
| 25 | ♀ ad. | May 9, 1873 | Henshaw | 3.90 | 3.27 | 0.60 | 0.71 |
| 98 | ♀ ad. | May 17, 1873 | ....do | 3.95 | 3.23 | 0.57 | 0.79 |

55. *Contophus virens* (L.), var. *richardsonii*, Sw.—Western Wood-Pewee.
Not seen till the 15th, when it was not common.

| No. | Sex. | Date. | Collector. | Wing. | Tail. | Bill. | Tarsus. |
|-----|------|-------|------------|-------|-------|-------|---------|
| 95 | ♂ ad. | May 17, 1873 | Henshaw | 3.41 | 2.69 | 0.54 | 0.51 |
| 276 | ♂ ad. | ............ | ............ | ........ | ........ | ........ | ........ |

56. *Empidonax minimus*, Bd.—Least Flycatcher.
Made its appearance the 12th. Apparently not common.

| No. | Sex. | Date. | Collector. | Wing. | Tail. | Bill. | Tarsus. |
|-----|------|-------|------------|-------|-------|-------|---------|
| 97 | ♂ ad. | May 17, 1873 | Henshaw | 2.58 | 2.43 | 0.40 | 0.41 |

57. *Empidonax obscurus* (Sw.)—Wright's Flycatcher.
First seen the 17th. Afterward rather common, keeping among the low trees and bushes.

| No. | Sex. | Date. | Collector. | Wing. | Tail. | Bill. | Tarsus |
|-----|------|-------|------------|-------|-------|-------|--------|
| 56 | ♂ ad. | May 12, 1873 | Henshaw | 2.79 | 2.75 | 0.47 | 0.69 |
| 94 | ♀ ad. | May 17, 1873 | ....do | 2.84 | 2.75 | 0.48 | 0.71 |

CAPRIMULGIDÆ (the Goatsuckers).

58. *Antrostomus nuttalli,* Aud.—Nuttall's Poor-Will.

May 13, a male was started from among some bushes. Two days later a second was taken in a similar locality. This may, perhaps, be taken as the time of the general arrival.

| No. | Sex. | Date. | Collector. | Wing. | Tail. | Bill. | Tarsus. |
|-----|------|-------|------------|-------|-------|-------|---------|
| 90 | ♂ ad. | May 15, 1873 | Henshaw .... ..... | 5.64 | 3.61 | 0.43 | 0.71 |

PICIDÆ (the Woodpeckers).

59. *Colaptes auratus* (L.), var. *mexicanus,* Sw.—Red-shafted Flicker.

Common in the cottonwood-groves along the Platte.

60. *Melanerpes erythrocephalus,* (L.)—Red-headed Woodpecker.

A single individual was taken by Dr. Rothrock, June 3, and was the only one seen.

| No. | Sex. | Date. | Collector. | Wing. | Tail. | Bill. | Tarsus. |
|-----|------|-------|------------|-------|-------|-------|---------|
| .... | ♂ ad. | June 3, 1873 | Rothrock .......... | 5.66 | 3.66 | 1.08 | 0.90 |

FALCONIDÆ (the Falcons).

61. *Falco saker,* Schleg., var. *polyagrus,* Cass.—Prairie-Falcon.

Of this falcon, a single male in adult plumage was shot. The specimen is of interest, as being the third only known to have been taken in this plumage. Mr. Ridgway has kindly compared this with the others in the Smithsonian collection, and finds it to present the following differences: The transverse bars of the upper surface are more sharply defined, and are pale earth-brown or dull ochraceous, instead of ashy-drab, and are very distinct on the rump instead of being entirely obsolete. *The upper parts lack entirely any bluish tinge,* which is so strongly marked on the other two specimens. The markings on the flanks are in the form of large transverse spots of dark vandyke-brown, with intervening rounded spots of pale reddish-drab. Cere, legs, and feet light yellow.

| No. | Sex. | Date. | Collector. | Wing. | Tail. | Bill. | Tarsus. |
|-----|------|-------|------------|-------|-------|-------|---------|
| 64 | ♂ ad. | May 12, 1873 | Henshaw .......... | 12.25 | 8.00 | 0.75 | 2.00 |

62. *Falco sparverius,* L.—Sparrow-Hawk.

Very common everywhere.

| No. | Sex. | Date. | Collector. | Wing. | Tail. | Bill. | Tarsus. |
|-----|------|-------|------------|-------|-------|-------|---------|
| 34 | ♂ ad. | May 9, 1873 | Henshaw .......... | 7.67 | 5.05 | 0.50 | 1.37 |

63. *Nisus fuscus*, Gm.—Sharp-shinned Hawk.
Rather common.  Creates sad havoc among the Turtle-Doves (*Zenaidura carolinensis*).

| No. | Sex. | Date. | Collector. | Wing. | Tail. | Bill. | Tarsus. |
|-----|------|-------|-----------|-------|-------|-------|---------|
| 35  | ♀ jun. | May 9, 1873 | Henshaw .......... | 7.70 | 6.75 | 0.50 | 2.00 |

COLUMBIDÆ (the Doves).

64. *Zenaidura carolinensis*, L.—Carolina Dove.
In very large numbers in the cottonwood-groves along the banks of the Platte, and elsewhere abundant.  The first nest was found on the ground May 7, and contained a freshly-laid egg.  This species is singularly indifferent in the choice of a location for its nest.  A favorite site is the thick undergrowth which clothes the trunks of the cottonwoods.  But nests may often be found in the same piece of woods, placed in bushes and on the ground; and in the latter case not infrequently in an entirely open place.  The nests are usually but a slight mass of straws and twigs irregularly disposed, and so slight is the structure that the eggs are often visible from the ground through the interstices.

| No. | Sex. | Date. | Collector. | Wing. | Tail. | Bill. | Tarsus. |
|-----|------|-------|-----------|-------|-------|-------|---------|
| 6   | ♂ ad. | June —, 1873 | Rothrock .......... | 5.53 | 7.14 | 0.58 | 0.75 |

CHARADRIIDÆ (the Plovers).

65. *Ægialitis vociferus*, L.
Abundant.  Breeds on the sandy shores of the Platte River in June. Deposits its eggs in a slight hollow in the sand.

| No. | Date. | Collector. | Wing. | Tail. | Bill. | Tarsus. |
|-----|-------|-----------|-------|-------|-------|---------|
| 14  | June — | Dr. Rothrock............ | 6.06 | 3.75 | 0.80 | 1.31 |

SCOLOPOCIDÆ (the Snipes).

66. *Gallinago gallinaria* (Gm.), var. *wilsonii*, Temm.—Wilson's Snipe.
Quite a number seen in marshy spots about small ponds.

67. *Macrorhamphus griseus* (Gm.)—Red-breasted Snipe.
An abundant migrant.

68. *Totanus melanoleucus* (Gm.)—Greater Yellow-Legs.
Numbers seen in the market.

69. *Totanus flavipes* (Gm.)—Lesser Yellow-Legs.
A few seen.

70. *Totanus solitarius* (Wils.)—Solitary Sandpiper.
Quite common about the ponds and along the streams.

71. *Tringoides macularius* (L.)—Spotted Sandpiper.
A few seen.

RECURVIROSTRIDÆ (the Avocets).

72. *Recurvirostra americana*, Gm.—Avocet.
Abundant. Numbers of this and the succeeding species find their way into the markets.

73. *Himantopus nigricollis*, Vieill.—Black-necked Stilt.
Less common than the preceding. Both frequent the small ponds and marshes.

RALLIDÆ (the Rails).

74. *Rallus virginianus*, L.—Virginia Rail.
A single one of this species was found skulking in a bed of rushes in early May.

ANATIDÆ (the Ducks).

75. *Anas boschas*, L.—Mallard.
Common.

76. *Nettion carolinensis*, Gm.—Green-winged Teal.
77. *Querquedula discors*, Steph.—Blue-winged Teal.
78. *Querquedula cyanoptera*, Cass.—Red-breasted Teal.
All abundant.

79. *Spatula clypeata* (L.)—Shoveler.
One of the most abundant of the family. Found in every pond and slough.

80. *Mareca americana* (Gm.)—American Widgeon.
Rather numerous.

81. *Aythya ferina* (L.), var. *americana*, Eyton.—Red-head.
A few seen.

PODICIPIDÆ (the Grebes).

82. *Podiceps auritus*, L., var. *californicus*, Heerm.—Eared Grebe.
Numerous in the ponds as late as the 15th.


SECTION II.

Fort Garland is situated in Southern Colorado, on the lowest bench of the Sierra Blanca Mountains, distant twenty miles east of the Rio Grande, in latitude 37° 25' north, longitude 105° 26' west, and has an elevation of 7,600 feet above the level of the sea. Immediately surrounding the post is a sage-brush plain, which to the northward and

westward stretches away for many miles, presenting the same unvary-
ing characteristics, but to the north and east is broken up by volcanic
ridges, which are soon lost in the foot-hills of the mountains. The foot-
hills are well clothed with piñons and cedars. From May 24 till June 3
the time was spent in making collections in the immediate vicinity of
the fort, more particularly upon the creeks which flow through the plain
and are well timbered with cottonwoods, and in many places skirted by
heavy brush. As might be expected, the immediate neighborhood of
these streams affords a home for large numbers of birds ; the number of
species, however, not being great, and of these by far the larger part
are of the smaller insectivorous kinds. The almost total absence of the
large rapacious birds was very noticeable, and during my whole stay in
the region I saw but two (*Buteo calurus* and *Buteo proticus*). A week's
camp in the pine-woods at the base of Mount Baldy, some twelve miles
to the north of the fort, at an approximate elevation of 9,500 feet, added
numerous varieties to the list, many of which were not met with at all
farther down, and also afforded an opportunity of observing the vertical
range of many of the species. The timber consists mainly of the yellow pine, which here attains a large size, interspersed with more or less
spruce.  Of the deciduous trees the aspens were the only numerous
representatives ; these grew in thick groves on slopes of the mountains,
and often attain a great elevation, sometimes, indeed, forming the tim-
ber-limit above the pine. The small streams are thickly skirted with
many deciduous bushes and shrubs, prominent among which are the
willows and alders. The fauna at this point is analogous to the Cana-
dian. On returning to Garland I was afforded an opportunity of making
a week's trip to the summer cavalry-camp established on the banks of
the Rio Grande, ninety miles northeast of the fort. Here I was most
kindly received by Captain Carraher and Lieutenant Pond, officers in
charge, who extended to me every courtesy and aid. The number and
variety of the birds found along the Rio Grande at this point did not
differ in any noteworthy respect from those in the vicinity of Garland,
and the collecting trips made into the mountains, which rise a few miles
from the banks, gave similar results to those obtained at Mount Baldy.
Returning to the post June 19, a short trip was made to a series of alkali
lakes, thirty miles northwest, and some interesting facts obtained regarding the nidification of the water-birds. The remaining time, till July 2, was
occupied in making daily excursions from the fort.  In conclusion I cannot refrain from mentioning the uniform courtesy I received from each
and all the officers of the post.  To Colonel Alexander, the commanding
officer, to Captain Jewitt, and to Lieutenant Hartz, whose hospitality I
enjoyed during my stay at the post, I am greatly indebted.  Every possible aid in the prosecution of my work was extended.

TURDIDÆ (the Thrushes).

1. *Turdus migratorius*, L.—Robin.
    Moderately common in this locality ; nests in the cottonwoods along
the streams.  A number of nests were examined, which were composed
largely of sheep's wool, which the birds find clinging to the bushes.

| No. | Sex. | Locality. | Date. | Collector. | Wing. | Tail. | Bill. | Tarsus. |
|-----|------|-----------|-------|-----------|-------|-------|-------|---------|
| 64 | ♀ ad. | South Park, Col ...... | June 27 | Rothrock . | 5.32 | 4.28 | 0.83 | 1.24 |

2. *Turdus pallasii*, Cab., var. *audubonii*, Bd.—Audubon's Thrush.

Abundant, especially in the aspen-groves along the mountain-sides, at an elevation of 10,000 feet, where in early morning I have heard not less than eight males singing in concert. During the breeding-season seldom seen lower than 8,000 feet. A nest found June 7 was built in the cavity of a broken pine-stub about three feet from the ground; was composed almost wholly of strips of bark and coarse grasses covered externally with mosses. It contained a single light-blue egg.

| No. | Sex. | Locality. | Date. | Collector. | Wing. | Tail. | Bill. | Tarsus. |
|---|---|---|---|---|---|---|---|---|
| 205 | ♂ ad. | Near Garland, Col .... | May 30 | Henshaw . | 3. 85 | 3. 00 | ...... | 1. 20 |
| 287 | ♀ ad. | ......do .............. | June 7 | .... do .... | 3. 98 | 2. 95 | 0. 50 | 1. 09 |
| 288 | ♂ ad. | ......do .............. | June 7 | .... do .... | 3. 75 | 3. 04 | 0. 54 | 1. 12 |

3. *Turdus fuscescens* Steph.—Tawny Thrush.

This species was found rather numerous along the streams below an elevation of about 8,000 feet. Two nests were found, both built on the ground. One taken June 19 and containing four freshly-laid eggs was curiously enough built in and over the nest of the previous year, the two making a pile some five inches high. Eggs blue, slightly darker than those of the preceding.

| No. | Sex. | Locality. | Date. | Collector. | Wing. | Tail. | Bill. | Tarsus. |
|---|---|---|---|---|---|---|---|---|
| 142 | ♂ ad. | Garland, Col.......... .... | May 26 | Henshaw . | 4. 05 | 3. 28 | 0. 60 | 1. 07 |
| 170 | ♂ ad. | ......do ........... .... | May 28 | .... do .... | 4. 19 | 3. 29 | 0. 60 | 1. 16 |
| 376 | ♀ ad. | ......do ,........ ....| June 19 | .... do .... | 3. 94 | 3. 11 | 0. 57 | 1. 07 |
| 577 | ♂ ad. | ......do ......... .... | June 19 | .... do .... | 4. 07 | 3. 13 | 0. 57 | 1. 12 |

4. *Galeoscoptes carolinensis* (L.)—Catbird.

In the thickets lining the small streams a few pairs of Catbirds were found. Their nests, built in low bushes and containing freshly-laid eggs, were taken about the middle of June.

5. *Oreoscoptes montanus* (Towns.)—Sage-Thrasher.

An occasional pair of these birds were noticed in the sage-brush in the immediate vicinity of Fort Garland, and young just from the nest were met with June 20. At the alkali lakes above mentioned a nest was discovered June 22, containing four eggs far advanced in incubation. The nest, a bulky structure of twigs lined with grass, was placed in a bush some three feet from the ground. Eight or ten inches above the nest was placed a platform of twigs, which, whatever may have been the original intention, certainly served as an admirable screen from the rays of an almost tropic sun. It possibly may have been intended as the site of the nest, and then for some reason have been abandoned for the one beneath.

| No. | Sex. | Locality. | Date. | Collector. | Wing. | Tail. | Bill | Tarsus. |
|---|---|---|---|---|---|---|---|---|
| 401 | ♂ ad. | Alkali lakes, Col ...... | June 22 | Henshaw . | 3. 98 | 3. 83 | 0. 77 | 1. 17 |

SAXICOLIDÆ (the Saxicolas).

6. *Sialia arctica*, Swains.—Arctic Bluebird.

Found common and breeding everywhere in this region from 7,000 feet upward. Nests frequently in a deserted woodpecker's hole, or as often in the natural cavity of some decayed stub. Two broods are reared during the season.

| No. | Sex. | Locality. | Date. | Wing. | Tail. | Bill. | Tarsus. |
|---|---|---|---|---|---|---|---|
| 143 | ♀ ad. | Garland, Col.............. | May 26 | 4.30 | 2.87 | 0.83 | 0.81 |
| 169 | ♂ ad. | ......do ................ | May 28 | 4.50 | 2.88 | 0.57 | 0.84 |
| 280 | ♀ ad. | ......do ................ | June 6 | 4.36 | 2.76 | 0.55 | 0.80 |
| 326 | ♂ ad. | Rio Grande, Col ......... | June 12 | 4.50 | 2.92 | 0.56 | 0.80 |
| 330 | ♂ ad. | ......do ................ | June 12 | 4.65 | 3.02 | 0.57 | 0.86 |
| 356 | ♂ ad. | ......do ................ | June 15 | 4.63 | 2.90 | 0.57 | 0.82 |
| 47 | ♂ ad. | South Park, Col ......... | June 25 | 4.35 | 2.91 | 0.61 | 0.85 |

SYLVIIDÆ (the Sylvias).

7. *Regulus calendula* (L.)

Very common in the heavy pine-forests at an elevation of 10,000 feet. Its song, for so diminutive a bird, is remarkably loud and clear, and no less wonderful for its sweetness and modulation. June 11, while collecting on a mountain near the Rio Grande, I discovered a nearly finished nest, built on a low branch of a pine, which I have little doubt belonged to this bird. The male was singing directly overhead, but although I watched for some time in hopes of being able to see the female in the act of building, I was disappointed. The nest was a somewhat bulky structure, very large for the size of the bird, externally composed of strips of bark, and lined thickly with feathers of the Grouse (*Canace obscura*).

| No. | Sex. | Locality. | Date. | Collector. | Wing. | Tail. | Bill. | Tarsus. |
|---|---|---|---|---|---|---|---|---|
| 215 | ♂ ad. | Near Garland, Col .... | May 30 | Henshaw. | 2.42 | 1.60 | 0.37 | 0.65 |
| 265 | ♂ ad. | ......do ................ | June 5 | ....do .... | 2.43 | 1.97 | 0.34 | 0.72 |

PARIDÆ (the Titmice).

8. *Parus atricapillus* (L.), var. *septentrionalis*, Harris.—Long-tailed Chickadee.

By no means as common here as the following species. Found indifferently in the heavy pine-woods and among the cottonwoods of the streams. Could detect no differences in habits and notes from the eastern Chickadee (*atricapillus*), from which it chiefly differs in its longer tail and lighter colors.

9. *Parus montanus*, Gambel.—Mountain-Chickadee.

Abundant. At this, the breeding-season, a rather exclusive inhabitant of the pine-woods. Like the preceding, a very active and persistent insect-hunter, exploring every crack and crevice beneath the rough bark for the hidden larvæ, which are instantly dragged forth, and, after being vigorously hammered on some horizontal limb and reduced to a shapeless mass, eagerly swallowed.

SITTIDÆ (the Nuthatches).

10. *Sitta carolinensis*, Gm., var. *aculeata*, Cass.—Slender-billed Nuthatch.

Common; chiefly puricoline in its habits, which correspond with those of its eastern representative.

| No. | Sex. | Locality. | Date. | Collector. | Wing. | Tail. | Bill. | Tarsus. |
|---|---|---|---|---|---|---|---|---|
| 226 | ♂ ad. | Garland, Col.......... | June 3 | Henshaw. | 3.45 | 2.04 | 0.75 | 0.65 |

11. *Sitta canadensis*, L.—Red-bellied Nuthatch.

Rather common and found only in the pine-woods. The most active and restless of the family. A nest of this species was found in a small pine-stub a few feet from the ground. The hole was excavated to the depth of five inches and thoroughly lined at bottom with fine shreds of pine-bark. The eggs, five in number, were far advanced toward hatching; color grayish-white, thinly spotted with reddish dots, which are confluent at the larger end.

| No. | Sex. | Locality. | Date. | Collector. | Wing. | Tail. | Bill. | Tarsus. |
|---|---|---|---|---|---|---|---|---|
| 211 | ♂ ad. | Near Garland, Col .... | May 30 | Henshaw. | 2.53 | 1.53 | 0.54 | 0.62 |
| 226 | ♀ ad. | ......do ............. | June 3 | ....do .... | 2.48 | 1.48 | 0.58 | 0.58 |

12. *Sitta pusilla*, Lath., var. *pygmæa*, Vigors.—Pigmy Nuthatch.

By June 12 I noticed these birds flying about the high pine-stubs, with food in their bills for their young. This nuthatch is very sociable in its habits, and is almost always found in large flocks. Even during the breeding-season several are usually to be seen in company, and it is not unusual at this time to find them associated with both the preceding species, and also the Titmice, the whole band apparently being on the best of terms with each other.

| No. | Sex. | Locality. | Date. | Collector. | Wing. | Tail. | Bill. | Tarsus. |
|---|---|---|---|---|---|---|---|---|
| 244 | ♂ ad. | Mountains near Garland, Col. | June 4 | Henshaw. | 2.44 | 1.58 | 0.56 | 0.60 |
| 283 | ♀ ad. | ......do............... | June 6 | .... do .... | 2.52 | 1.57 | 0.54 | 0.54 |

CERTHIADÆ (the Creepers).

13. *Certhia familiaris*, L., var. *americana*, Bon.—Brown Creeper.

Of rather frequent occurrence in the pine-region. Several specimens were obtained, which were evidently breeding.

| No. | Sex. | Locality. | Date. | Collector. | Wing. | Tail. | Bill. | Tarsus. |
|---|---|---|---|---|---|---|---|---|
| 227 | ♂ ad. | Near Garland, Col ..... | June 3 | Henshaw. | 2.59 | 2.14 | 0.69 | 0.55 |

TROGLODYTIDÆ (the Wrens).

14. *Troglodytes aëdon*, Vieill., var. *parkmanni*, Aud.—Parkman's Wren.
Very abundant, inhabiting the undergrowth of the streams. I found a pair building May 23. Nearly a month later a nest was obtained, built in a small stub. The hole was nearly filled up with a mass of twigs, in the center of which was left a deep cavity, lined with sheep's wool and feathers. It contained but a single egg. This was white, covered with fine reddish-brown spots.

| No. | Sex. | Locality. | Date. | Collector. | Wing. | Tail. | Bill. | Tarsus. |
|-----|------|-----------|-------|------------|-------|-------|-------|---------|
| 68 | ♂ ad. | South Park, Col ....... | June 27 | Rothrock. | 2.06 | 1.90 | 0.54 | 0.72 |

15. *Telmatodytes palustris* (Wils.), *paludicola*, Baird.—Long-billed Marsh-Wren. .
This species was numerous among the reeds of the alkali ponds in this vicinity. June 23, they were apparently just laying, as two nests were found, each containing but one egg. This species is one of the very few that seems never to vary in the modeling of its curious nest. These are always nearly spherical balls of tightly-woven rushes, a small hole being left in the side as an entrance, and thickly lined inside with down and feathers.

| No. | Sex. | Locality. | Date. | Collector. | Wing. | Tail. | Bill. | Tarsus. |
|-----|------|-----------|-------|------------|-------|-------|-------|---------|
| 167 | ♀ ad. | Alkali lakes, Col ...... | May 28 | Henshaw. | 2.03 | 1.94 | 0.50 | 0.66 |
| 178 | ♂ ad. | ......do ............... | May 29 | ....do .... | 2.00 | 2.00 | ...... | 0.67 |
| 241 | ♂ ad. | ......do ............... | June 4 | ....do .... | 2.04 | 1.95 | 0.51 | 0.66 |

SYLVICOLIDÆ (the Warblers).

16. *Helminthophaga celata* (Say.)—Orange-crowned Warbler.
Met with but on one or two occasions. Shows a preference at this season for the scrub-covered mountain-sides. A male was taken in an aspen-grove at an elevation of about 11,000 feet. The song is short, but spirited, and consists of a few trills, ending with a rising inflection.

| No. | Sex. | Locality. | Date. | Collector. | Wing. | Tail. | Bill. | Tarsus. |
|-----|------|-----------|-------|------------|-------|-------|-------|---------|
| 291 | ♂ ad. | Near Garland, Col .... | May 9 | Henshaw. | 2.55 | 2.07 | 0.45 | 0.72 |

17. *Dendroica æstiva* (Gm.)—Yellow Warbler.
Not uncommon among the deciduous trees of the streams. Several nests placed in bushes were obtained, and showed a general similarity in structure to the usual style. One, however, made of sheep's wool and hempen material, lined with fine grasses and feathers, has more the appearance of a flycatcher's nest. Except that it is thicker and more carefully made, it might be mistaken for that of *Empidonax pusillus*. The ground-color of the eggs taken in the West is pure white, and lacks the greenish tinge which is characteristic of all eastern specimens I have ever seen.

18. *Dendroica audubonii* (Towns.)—Audubon's Warbler.

This species was a moderately common one in the pine-region from about 9,000 feet upward. By the 1st of June all were paired, and on the 3d I saw a female just beginning a nest in the top of a small spruce, some thirty feet from the ground. This was finished June 8, but no eggs had been laid, and I was forced to content myself with the nest alone. Outwardly it was composed of strips of bark firmly and neatly woven, and lined with fine grasses. It has an external diameter of four inches and is one inch deep.

| No. | Sex. | Locality. | Date. | Collector. | Wing. | Tail. | Bill. | Tarsus. |
|---|---|---|---|---|---|---|---|---|
| 290 | ♂ ad. | Near Garland, Col | June 6 | Henshaw. | 3.23 | 2.54 | 0.43 | 0.74 |
| 357 | ♂ ad. | ......do | June 7 | .... do .... | 3.11 | 2.40 | 0.41 | 0.70 |

19. *Dendroica nigrescens* (Towns.)—Black-throated Gray Warbler.

A warbler was seen June 25 in a grove of pine-trees on the sides of a narrow cañon, which I am quite confident was of this species. It had a short, feeble, but rather pleasing song, which it constantly emitted at short intervals as it flew from tree to tree. Owing to its shyness, I did not succeed in capturing it.

20. *Geothlypis macgillivrayi* (Aud.)—Macgillivray's Warbler.

Somewhat common along the streams; not observed at a higher altitude than 9,000 feet; keeps much in the swampy thickets, where it searches industriously under fallen logs and among the dead leaves for insects, occasionally pausing in its labors to warble its short pleasing song.

| No. | Sex. | Locality. | Date. | Collector. | Wing. | Tail. | Bill. | Tarsus. |
|---|---|---|---|---|---|---|---|---|
| 130 | ♀ ad. | Garland, Col | May 25 | Henshaw. | 2.38 | 2.47 | 0.45 | 0.80 |
| 160 | ♂ ad. | ......do | May 28 | .... do .... | 2.57 | 2.45 | 0.42 | 0.81 |

21. *Myiodioctes pusillus* (Wils.)—Green Black-capped Flycatcher.

During the last days of May a few stragglers were seen among the cottonwoods, apparently still on their way northward. Perhaps a few remain to breed.

| No. | Sex. | Locality. | Date. | Collector. | Wing. | Tail. | Bill. | Tarsus. |
|---|---|---|---|---|---|---|---|---|
| 44 | ♂ ad. | South Park, Col | June 24 | Rothrock. | 2.23 | 2.17 | 0.39 | 0.73 |
| 161 | ♀ ad. | Garland, Col | May 28 | Henshaw. | 2.17 | 2.18 | 0.40 | 0.65 |
| 162 | ♂ ad. | ......do | May 28 | .... do .... | 2.12 | 2.19 | 0.35 | 0.68 |
| 163 | ♂ ad. | ......do | May 28 | .... do .... | 2.25 | 2.07 | 0.38 | 0.68 |

22. *Setophaga ruticilla* (L.)—Redstart.

Not met with at Garland, nor earlier at Denver, but numbers seen May 23 on the Huerfano River, eighty miles northeast of Garland.

HIRUNDINIDÆ (the Swallows).

23. *Petrochelidon lunifrons* (Say.)—Cliff-Swallow.

In large numbers, building under the eaves of the post-quarters. I noticed here a very curious departure from the usual method of constructing the nest. Under the projecting eaves of one of the store-houses a large colony had established themselves, there being in the neighborhood of fifty nests, most of which were built in the usual fashion. But a few pairs, taking advantage of circumstances, had established themselves in certain small passages which opened directly under the eaves, and had served as ventilators. The mouth of each one of these had been built up with mud, a small hole being left as an entrance. Some twelve inches beyond was the nest proper, consisting of a small pile of straws and feathers, on which the eggs were deposited. The wisdom of the birds in thus availing themselves of these holes was very clearly demonstrated, since nearly the entire labor of nest-making was obviated and a much safer domicile secured.

| No. | Sex. | Locality. | Date. | Collector. | Wing. | Tail. | Bill. | Tarsus. |
|-----|------|-----------|-------|------------|-------|-------|-------|---------|
| 422 | ♀ ad. | Garland, Col .......... | June 27 | Henshaw . | 4. 28 | 2. 13 | 0. 30 | 0. 45 |

24. *Hirundo horreorum*, Barton.—Barn-Swallow.

Common, nesting in the stables and out-buildings of the post.

| No. | Sex. | Locality. | Date. | Collector. | Wing. | Tail. | Bill. | Tarsus. |
|-----|------|-----------|-------|------------|-------|-------|-------|---------|
| 66 | ...... | South Park, Col. ....... | June 26 | Rothrock . | 4. 73 | 4. 07 | 0. 28 | 0. 42 |
| 161 | ♂ jun. | Twin Lakes, Col ...... | Aug. — | .... do .... | 4. 37 | 3. 10 | 0. 32 | 0. 43 |

25. *Tachycineta bicolor* (Vieill.)—White-bellied Swallow.

Not uncommon; nests here in old stubs. Not seen at a higher altitude than 8,000 feet.

| No. | Sex. | Locality. | Date. | Collector. | Wing. | Tail. | Bill. | Tarsus. |
|-----|------|-----------|-------|------------|-------|-------|-------|---------|
| 83 | ♀ ad. | South Park, Col ....... | July 1 | Rothrock . | 4. 55 | 2. 50 | 0. 25 | 0. 47 |

26. *Tachycineta thalassina* (Sw.)—Violet-green Swallow.

A few pairs were seen during the breeding-season in the same localities as the White-bellied Swallow. This species, however, attains a much higher altitude, and at 10,000 feet I found it very common and in large colonies. June 7, they had not begun to build, though evidently about to do so in the high pine-stubs.

| No. | Sex. | Locality. | Date. | Collector. | Wing. | Tail. | Bill. | Tarsus. |
|-----|------|-----------|-------|------------|-------|-------|-------|---------|
| 254 | ♂ ad. | Mountains near Gar-land, Col. | June 5 | Henshaw . | 4. 60 | 1. 98 | 0. 23 | 0. 41 |
| 255 | ♂ ad. | ......do ............... | June 5 | .... do .... | 4. 55 | 2. 06 | 0. 22 | 0. 42 |
| 256 | ♂ ad. | ......do ............... | June 5 | .... do .... | 4. 51 | 2. 07 | 0. 23 | 0. 40 |
| 257 | ♂ ad. | ......do ............... | June 5 | .... do .... | 4. 45 | 2. 05 | 0. 25 | 0. 40 |
| 258 | ♂ ad. | ......do ............... | June 5 | .... do .... | 4. 70 | 2. 15 | 0. 21 | 0. 37 |
| 259 | ♀ ad. | ......do ............... | June 5 | .... do .... | 4. 45 | 1. 95 | 0. 22 | 0. 42 |
| 260 | ♀ ad. | ......do ............... | June 5 | .... do .... | 4. 20 | 2. 00 | 0. 20 | 0. 44 |
| 279 | ♀ ad. | ......do ............... | June 6 | .... do .... | 4. 22 | 1. 90 | 0. 20 | 0. 43 |

27. *Stelgidopteryx serripennis*, (Aud.)—Rough-winged Swallow.

Not uncommon along the streams in the immediate vicinity of the post. I think they must have nested in the hollows of trees, since I could discover no bank suitable for their excavations.

## VIREONIDÆ (the Vireos).

28. *Vireo gilvus* (Vieill.), var. *swainsoni*, Bd.—Western Warbling Vireo.

Very common in the cottonwoods along the streams. Have observed it from 6,000 to 10,000 feet, at which latter height it is very numerous. There is nothing in habits and notes to distinguish it from the closely-allied eastern form (*gilvus*). It is, however, easily recognizable by its paler colors.

| No. | Sex. | Locality. | Date. | Collector. | Wing. | Tail. | Bill. | Tarsus. |
|-----|------|-----------|-------|------------|-------|-------|-------|---------|
| 138 | ♂ ad. | Near Garland, Col..... | May 26 | Henshaw. | 2.83 | 2.27 | 0.43 | 0.68 |
| 175 | ♂ ad. | ......do ............... | May 29 | .... do .... | 2.73 | 2.27 | 0.45 | 0.66 |
| 195 | ♀ ad. | ......do ............... | May 29 | .... do .... | 2.60 | 2.20 | 0.43 | 0.66 |
| 196 | ♂ ad. | ......do ............... | May 30 | .... do .... | 2.80 | 2.25 | 0.43 | 0.69 |
| 231 | ♀ ad. | ......do ............... | May 3 | .... do .... | 2.76 | 2.16 | 0.42 | 0.70 |
| 232 | ♂ ad. | ......do ............... | June 5 | .... do .... | 2.85 | 2.30 | 0.40 | 0.70 |
| 278 | ♀ ad. | ......do ............... | June 6 | .... do .... | 2.75 | 2.17 | 0.42 | 0.70 |
| 380 | ♂ ad. | ......do ............... | June 19 | .... do .... | 2.84 | 2.19 | 0.42 | 0.69 |
| 389 | ♂ ad. | ......do ............... | June 21 | .... do .... | 2.83 | 2.27 | 0.43 | 0.68 |

29. *Vireo solitarius* (Wils.), var. *plumbeus*, Cs.—Lead-colored Vireo.

Not very common; frequenting about the same localities as the preceding, with perhaps a greater preference for the pine-woods; habits and song identical with the Solitary Vireo of the East. A nest found by Mr. C. E. Aiken in El Paso County, Colorado, and by him kindly presented to me, exhibits but little difference when compared with nests of the true *solitarius* taken in New England. It is composed of soft, cottony substances, bound externally with strips of bark and other fibrous material, with a lining of fine grasses. The eggs are pure white, spotted chiefly at the larger end with reddish-brown.

## AMPELIDÆ (the Chatterers).

30. *Myiadestes townsendii* (Aud.)—Townsend's Solitaire.

During a week's stay at the base of Baldy Peak I frequently saw this bird in the pine-forests, and as high up on the mountain-sides as 10,000 feet, and its summer-range doubtless extends up to timber-line. Its habits, as far as I noticed them, are singularly like those of the blue-birds. Besides a loud, liquid call-note, the male has a beautiful warbling song. This somewhat resembles the finest efforts of the Purple Finch (*Carpodacus purpureus*), but far excels that bird in power, sweetness, and modulation. Though I searched most carefully for the nest of this species, I was not successful further than to satisfy myself that it breeds in the crevices of the rocks. Its preference for such localities during the summer, with the evident solicitude manifested on more than one occasion, left little doubt in my mind upon this point.

| No. | Sex. | Locality. | Date. | Collector. | Wing. | Tail. | Bill. | Tarsus. |
|---|---|---|---|---|---|---|---|---|
| 277 | ♂ ad. | Garland, Col........... | June 6 | Henshaw . | 4.76 | 4.44 | 0.56 | 0.82 |
| 286 | ♂ ad. | Rio Grande, Col....... | June 7 | .... do .... | 4.64 | 4.31 | 0.54 | 0.75 |
| 324 | ♀ ad. | ......do................. | June 11 | .... do .... | 4.66 | 4.17 | 0.54 | 0.80 |
| 325 | ♂ ad. | ......do................. | June 12 | .... do .... | 4.47 | 4.20 | 0.54 | 0.82 |

LANNIIDÆ (the Shripes).

31. *Collurio ludovicianus* (L.), var. *excubitoroides*, Sw.—White-rumped Shrike.

Apparently rather rare in this locality, as I saw but one or two.

TANAGRIDÆ (the Tanagers).

32. *Pyranga ludoviciana* (Wils.)—Louisiana Tanager.

Common, and nowhere more so than among the pines, at an elevation of 10,000 feet. Its song is fine, and not unlike that of the Scarlet Tanager (*P. rubra*).

| No. | Sex. | Locality. | Date. | Collector. | Wing. | Tail. | Bill. | Tarsus. |
|---|---|---|---|---|---|---|---|---|
| 192 | ♂, young of year. | Garland, Col....... | May 29 | Henshaw . | 3.77 | 3.04 | 0.61 | 0.77 |
| 193 | ♀ ad. | ......do ........... | May 29 | .... do .... | 3.66 | 3.04 | 0.66 | 0.78 |
| 206 | ♂ ad. | ......do ........... | May 30 | .... do .... | 3.85 | 3.05 | 0.61 | 0.76 |
| 253 | ♀ ad. | ......do ........... | June 5 | .... do .... | 3.58 | 2.90 | 0.63 | 0.76 |
| 281 | ♂ ad. | ......do . ........ | June 6 | .... do .... | 3.82 | 3.00 | 0.60 | 0.74 |
| 321 | ♂ ad. | Rio Grande, Col.... | June 12 | .... do .... | 3.83 | 3.08 | 0.60 | 0.75 |
| 322 | ♂ ad. | ......do . ,........ | June 12 | .... do .... | 3.82 | 2.97 | 0.81 | 0.79 |

FRINGILLIDÆ (the Finches).

33. *Carpodacus cassinii*, Bd.—Cassin's Purple Finch.

Apparently not common in this region. A single small flock was seen the middle of June near the Rio Grande. The song much resembled that of the eastern Purple Finch (*C. purpureus*).

34. *Carpodacus frontalis* (Say).—House-Finch; Burion.

Apparently rather rare in the vicinity of Garland; but at Taos, seventy-five miles farther south, I saw great numbers. A large colony had established their nests in the interstices of a thatched roof of a shed directly adjoining the house. These nests were bulky, inartistic structures, made of twigs and sheep's wool; eggs five in number, greenish-blue, spotted with black. From the extreme sociability of these birds and their beautiful song, they are great favorites, and are carefully protected in all the towns.

| No. | Sex. | Locality. | Date. | Collector. | Wing. | Tail. | Bill. | Tarsus. |
|---|---|---|---|---|---|---|---|---|
| 768 | ♂ ad. | Garland, Col.......... | May 28 | Henshaw . | 3.10 | 2.55 | 0.40 | 0.67 |

35. *Chrysomitris pinus* (Wils.)—Pine-Finch.

May 29, a flock of perhaps fifteen individuals were seen. These were in the breeding-dress, and the species undoubtedly spends the summer in the mountains of Colorado.

| No. | Sex. | Locality. | Date. | Collector. | Wing. | Tail. | Bill. | Tarsus. |
|---|---|---|---|---|---|---|---|---|
| 180 | ♂ ad. | Garland, Col.......... | May 29 | Henshaw. | 2.88 | 1.98 | 0.41 | 0.57 |
| 181 | ♀ ad. | ......do............. | May 29 | .... do .... | 2.73 | 1.82 | 0.40 | 0.55 |
| 182 | ♂ ad. | ......do............. | May 29 | .... do .... | 2.68 | 1.88 | 0.48 | 0.54 |

36. *Loxia curvirostra*, L., var. *americana*, Wils.—Red Crossbill.

Probably breeds in the pine-region. Several specimens were taken during the month of June that evidently had raised their young long before.

37. *Leucosticte australis*,'Allen.—Gray-crowned Finch.

The following interesting notes are given by Dr. Rothrock, who found the species very abundant in the mountains back of Fairplay, in the South Park, and also at Mounts Harvard, Evans, Red Mountains, and elsewhere. These birds are, in habitat, the associates of the White-tailed Ptarmigan, and, like that bird, are never found below the timber-line in summer, ranging thence upward to the summits of the highest peaks. It is never found singly, but usually in flocks of from six to thirty, and rarely far away from large bodies of snow; its favorite resort being the edges of snow-banks, where they find grass-seeds, and also a small black coleopterous insect. Even when found among the scrub-pines, which was rarely the case, it was noticed that they seldom alighted on a tree, but kept constantly on the ground. At all times they were rather shy and suspicious. The specimens taken were all in breeding-dress.

| No. | Sex. | Locality. | Date. | Collector. | Wing. | Tail. | Bill. | Tarsus. |
|---|---|---|---|---|---|---|---|---|
| 69 | ♂ ad. | South Park, Col...... | July 2 | Rothrock. | 3.98 | 3.00 | 0.75 | 0.38 |
| 69a | ♂ ad. | ......do............. | July 2 | .... do .... | 4.13 | 3.07 | 0.74 | 0.50 |
| 69b | ♂ ad. | ......do ............ | July 2 | .... do .... | 4.20 | 3.12 | 0.75 | 0.43 |
| 69c | ♂ ad. | ......do............. | July 2 | .... do .... | 4.23 | 3.13 | 0.75 | 0.42 |
| 69d | ♂ ad. | ......do............. | July 2 | .... do .... | 4.24 | 3.12 | 0.75 | 0.48 |

38. *Passerculus savanna*, Wils., var. *alaudinus*, Bon.—Western Savannah Sparrow.

Common in this region as elsewhere throughout the West. Generally found in the vicinity of water.

| No. | Sex. | Locality. | Date. | Collector. | Wing. | Tail. | Bill. | Tarsus. |
|---|---|---|---|---|---|---|---|---|
| 212 | ♂ ad. | Garland, Col.......... | May 30 | Henshaw. | 2.81 | 2.17 | 0.43 | 0.82 |
| 213 | ♂ ad. | ......do............. | May 30 | .... do .... | 2.79 | 2.14 | 0.43 | 0.75 |
| 214 | ♂ ad. | ......do............. | May 30 | .... do .... | 2.83 | 2.24 | 0.48 | 0.82 |

39. *Poöcœtes gramineus* (Gm.), var. *confinis*, Bd.—Western Grass-Finch. Abundant. Nests on the ground among the sage-brush. Two nests were obtained in South Park, Colorado, by Dr. Rothrock. Nest a slight structure of dried grasses, lined slightly with cottony substances from plants. Eggs four or five in number, of a greenish-white ground-color, blotched all over with light-brown and obsolete markings of purple, with a few black streakings.

40. *Zonotrichia leucophrys* (Forst.)—White-crowned Sparrow.

This species was not observed by me after the 1st of June. Dr. Rothrock, however, found it breeding in the South Park.

41. *Junco caniceps* (Woodh.)—Red-backed Snowbird.

In the heavy pine-woods in the neighborhood of Garland and among the bushes that fringe the small mountain-streams, this snowbird was the most abundant species of the locality. By the 1st of June the greater number appeared to be paired and breeding, though I was not able, after a careful search, to find their nests. The song consists of a rapid succession of low, trilling notes, which is usually emitted from the top of some low spruce or pine. Upon leaving the mountains of Colorado, this species was left behind, and in New Mexico is replaced in the mountains by the closely-allied form *dorsalis*.

| No. | Sex. | Locality. | Date. | Collector. | Wing. | Tail. | Bill. | Tarsus. |
|---|---|---|---|---|---|---|---|---|
| 207 | ♂ ad. | Mountains near Garland, Col. | May 30 | Henshaw. | 3.17 | 3.02 | 0.50 | 0.79 |
| 208 | ...... | ......do ............... | May 30 | .... do .... | 3.02 | 2.81 | 0.48 | 0.82 |
| 209 | ♀ ad. | ......do ............... | June 3 | .... do .... | 2.98 | 2.83 | 0.47 | 0.81 |
| 229 | ♂ ad. | ......do ............... | June 3 | .... do .... | 3.14 | 2.92 | 0.50 | 0.77 |
| 264 | ♂ ad. | ......do ............... | June 3 | .... do .... | 3.11 | 2.98 | 0.49 | 0.78 |
| 282 | ♀ ad. | ......do ............... | June 6 | .... do .... | 3.14 | 2.81 | 0.50 | 0.75 |
| 283 | ♂ ad. | ......do ............... | June 6 | .... do .... | 3.30 | 3.00 | 0.47 | 0.79 |
| 284 | ♂ ad. | ......do ............... | June 6 | .... do .... | 3.36 | 3.06 | 0.47 | 0.78 |
| 292 | ♀ | ......do ............... | June 7 | .... do .... | 2.92 | 2.71 | 0.47 | 0.80 |

42. *Spizella socialis* (Wils.)—Chipping Sparrow.

Not uncommon, keeping to the wooded districts along the streams. Notes and habits precisely as at the East.

| No. | Sex. | Locality. | Date. | Collector. | Wing. | Tail. | Bill. | Tarsus. |
|---|---|---|---|---|---|---|---|---|
| 355 | ♂ ad. | Rio Grande, Col........ | June 15 | Henshaw . | 2.85 | 2.62 | 0.37 | 0.65 |
| 423 | ♂ ad. | Garland, Col........... | June 27 | .... do .... | 2.88 | 2.50 | 0.39 | 0.70 |

43. *Spizella pallida* (Sw.), var. *breweri*, Cass.—Brewer's Sparrow.

Rather numerous. Inhabits the sage-brush and greasewood of the plains. Its song is short and weak, and somewhat resembles that of the Yellow-winged Sparrow (*C. passerinus*). It consists of a short prelude, followed by a succession of short, quickly-uttered notes, very well expressed by the striking together of pebbles.

| No. | Sex. | Locality. | Date. | Collector. | Wing. | Tail. | Bill. | Tarsus. |
|---|---|---|---|---|---|---|---|---|
| 141 | ♂ ad. | Fort Garland, Col..... | May 26 | Henshaw | 2.52 | 2.44 | 0.47 | 0.66 |
| 179 | ♂ ad. | ......do .............. | May 29 | .... do .... | 2.40 | 2.54 | 0.38 | 0.73 |
| 411 | ♂ ad. | ......do .............. | June 23 | .... do .... | 2.37 | 2.52 | 0.37 | 0.68 |

**44. *Melospiza melodia* (Wils.), var. *fallax*, Bd.—Western Song-Sparrow.**
Not common. Frequents moist localities in the neighborhood of water.

| No. | Sex. | Locality. | Date. | Collector. | Wing. | Tail. | Bill. | Tarsus. |
|---|---|---|---|---|---|---|---|---|
| 134 | ♂ ad. | Garland, Col.......... | May 25 | Henshaw . | 2.77 | 2.88 | 0.52 | 0.86 |
| 177 | ♂ ad. | ......do .............. | May 29 | .... do ... | 2.80 | 2.88 | 0.50 | 0.82 |
| 428 | ♀ ad. | ......do .............. | June 28 | .... do .... | 2.53 | 2.82 | 0.57 | 0.83 |

**45. *Melospiza lincolni* (Aud.)—Lincoln's Finch.**
By the last of May nearly all of this species had disappeared, migrating to the northward. A few stragglers were, however, seen here; shy and retiring in their habits, preferring the deep thickets along the streams. A few may possibly remain during the summer among the high mountains.

| No. | Sex. | Locality. | Date. | Wing. | Tail. | Bill. | Tarsus. |
|---|---|---|---|---|---|---|---|
| 133 | ♂ ad. | Garland, Col ..................... | May 25 | 2.45 | 2.50 | 0.49 | 0.79 |

**46. *Hedymeles melanocephalus* (Sw.)—Black-headed Grossbeak.**
Rather rare. But one or two seen.

| No. | Sex. | Locality. | Date. | Collector. | Wing. | Tail. | Bill. | Tarsus. |
|---|---|---|---|---|---|---|---|---|
| 135 | ♂ ad. | Garland, Col .......... | May 25 | Henshaw . | 3.93 | 3.15 | 0.70 | 0.91 |
| 373 | ♂ ad. | ......do .............. | June 19 | .... do .... | 4.27 | 3.69 | 0.72 | 0.88 |

**47. *Cyanospiza amœna* (Say).—Lazuli Finch.**
None seen in this vicinity. Common on the Huerfano River, sixty miles northeast of Garland. Song resembles in its character that of the Indigo-bird (*C. cyanea*), but is much weaker, and the strains less melodious.

**48. *Pipilo maculatus* (Sw.), var. *megalonyx*, Bd.—Long-spurred Towhee.**
Uncommon. A few were seen skulking among the dense undergrowth, and very shy.

**49. *Pipilo aberti*, Bd.—Abert's Towhee.**
Though no specimens were secured, pretty good evidence of the presence of this species at the alkali lakes northwest of Garland was

6 o s

obtained by the discovery of a nest containing two eggs, which a careful comparison with specimens in the Smithsonian Institution satisfies me must have belonged to this bird. It had evidently been deserted a short time before. The ground-color of the eggs is a faint bluish-white, with heavy black blotches and streaks at the larger end.

50. *Pipilo chlorurus* (Towns.)—Green-tailed Finch.

Common. Frequenting both the sage-brush of the plains and thickets of the streams. Nests both in bushes and on the ground. Nest composed of stalks of weeds and coarse grasses, lined with rootlets and fine grass. Eggs usually four or five in number, bluish-white, spotted with reddish-brown and purple. In one nest was found an egg of the Cow-Bunting.

### ICTERIDÆ (the Orioles).

51. *Dolichonyx oryzivorus* (L.)—Bobolink.

At the Huerfano crossing, May 22, three or four individuals were seen in company, apparently migrating. Not found at Garland.

52. *Molothrus pecoris* (Gm.)—Cow-Bunting.

An egg found in the nest of *Pipilo chlorurus* was the only indication of the presence of the species.

53. *Agelaius phœniceus* (L.)—Red-winged Blackbird.

Common in the marshes of the alkali lakes, breeding plentifully among the rushes.

54. *Xanthocephalus icterocephalus* (Bon.)—Yellow-headed Blackbird.

Very numerous at same locality as the preceding. June 22, many nests were found, some containing young, others fresh eggs, while others still were in process of construction. Always gregarious. Flocks of these birds were numerous on the Rio Grande in June.

55. *Sturnella magna* (L.), var. *neglecta*, Aud.—Western Meadow-Lark.

Numerous.

| No. | Sex. | Locality. | Date. | Collector. | Wing. | Tail. | Bill. | Tarsus. |
|-----|------|-----------|-------|------------|-------|-------|-------|---------|
| 194 | ♂ ad. | Garland, Col ......... | May 29 | Henshaw. | 4.90 | 3.40 | 1.30 | 1.40 |

56. *Icterus bullockii* (Sw.)—Bullock's Oriole.

But few individuals seen in this region. Song rich, clear, and melodious.

57. *Scolecophagus cyanocephalus* (Wagl.)—Brewer's Blackbird.

In large numbers on every creek and in each marshy spot. The resemblance of their actions and notes to those of the Rusty Blackbird (*S. ferrugineus*) have been noticed by all observers. In the choice of a nesting-site they are extremely variable, usually, however, building quite low. Have found their nests in trees, in bushes, in tussocks of grass, and beneath the overhanging banks of streams. These are six or seven inches in diameter, very bulky, and composed of sticks, weeds, and coarse grasses, inside which is usually a thick layer of mud, lined with rootlets and fine grasses. The eggs vary in number from four to six, usually five; the color varies from a dull olivaceous to a pale bluish-white, and are thickly covered with blotches of light-brown and burned-umber, this latter color often in the form of wavering lines. In some specimens the brown spots are confluent, and nearly hide the ground-color.

| No. | Sex. | Locality. | Date. | Collector. | Wing. | Tail. | Bill. | Tarsus. |
|-----|------|-----------|-------|-----------|-------|-------|-------|---------|
| 171 | ♂ ad. | Garland, Col.......... | May 28 | Henshaw. | 5. 47 | 4. 66 | 0. 90 | 1. 33 |
| 197 | ♂ ad. | ......do ............. | May 29 | ....do .... | 5. 08 | 4. 50 | 0. 90 | 1. 23 |
| 377 | ♀ ad. | ......do ............. | June 19 | ....do .... | 5. 13 | 4. 28 | 0. 86 | 1. 26 |
| 378 | ♂ ad. | ......do ............. | June 19 | ....do .... | 4. 73 | 3. 75 | 0. 77 | 1. 17 |
| 379 | ♀ ad. | ......do ............. | June 19 | ....do .... | 4. 47 | 3. 82 | 0. 85 | 1. 20 |
| 385 | ♂ ad. | ......do ............. | June 20 | ....do .... | 5. 30 | 4. 28 | 0. 86 | 1. 24 |
| 424 | ♂ ad. | ......do .....~... ... | June 28 | ....do .... | 5. 11 | 4. 21 | 0. 88 | 1. 15 |
| 425 | ♂ ad. | ......do ............. | June 28 | ....do .... | 5. 15 | 4. 10 | 0. 87 | 1. 22 |
| 426 | ♂ ad. | ......do ............. | June 28 | ....do .... | 4. 93 | 4. 25 | 0. 83 | 1. 26 |

CORVIDÆ (the Crows).

58. *Corvus corax*, L., var. *carnivorus*, Bart.—Raven.

Abundant; especially numerous about cattle-ranches. Breed usually on inaccessible cliffs.

59. *Picicorvus columbianus* (Wils.)—Nutcracker; Clarke's Crow.

During the latter part of May, I met with this species once or twice in the neighborhood of Baldy Peak, ten miles from Garland. They appeared very uneasy, flying about and alighting on the high pine-stubs, but their extreme shyness rendered it impossible to approach within satisfactory observing distance. As the previous year in Utah, where this was an abundant species, their shyness and habit of constantly moving from place to place made all attempts to even procure a speci-men fruitless, my surprise may be imagined when, on visiting the sum-mer cavalry-camp established on the Rio Grande, I found these birds regular daily visitors about camp, and exhibiting the same confiding familiarity as does the well-known Canada Jay or Whisky-Jack (*Peri-soreus canadensis*) of the north in the lumberman's camp. Early in the morning their well-known hoarse, rattling cries proclaimed their pres-ence, as they flew down from the tops of the high pine-clothed ridges, where at night they always retired to roost. So tame had they become that they would frequently alight on the ground, or the low branch of a tree, but a few feet distant from the lookers-on, and on one occasion a fearless individual was seen to enter a tent. On the ground, their motions appeared somewhat awkward, and they were only perfectly at home when among the pine-trees, in a small grove of which the tents were pitched. They eagerly seized upon any of the refuse thrown away by the cook, and scraps of meat were readily taken, and, if too large to be easily swallowed, carried up to the nearest horizontal limb and vigorously hammered till reduced to proper fragments. The corn and grain scattered about by the horses when feeding were also special objects of attention. They were rather quarrelsome, and when a con-tented croak betrayed the finder of some titbit a number instantly made a dive for the fortunate possessor, and both pursuers and pursued would disappear among the pines. I have little doubt but that they nest in the cavities of trees, and one was seen to enter a hole, which contained apparently the remains of an old nest. Young birds taken in June are easily distinguished from the old by the general hoariness of the plumage. In these the bluish ash is replaced to a great extent by a plumbeous white, becoming almost pure white about the throat.

| No. | Sex. | Locality. | Date. | Collector. | Wing. | Tail. | Bill. | Tarsus. |
|-----|------|-----------|-------|------------|-------|-------|-------|---------|
| 190 | ♂ ad. | Near Garland, Col .... | May 29 | Henshaw. | 7.25 | 4.92 | 1.66 | 1.38 |
| 309 | ♂ ad. | Rio Grande, Col ...... | June 10 | ....do .... | 7.75 | 4.81 | 1.65 | 1.40 |
| 310 | ♂ ad. | ......do ............... | June 10 | ....do .... | 7.37 | 4.87 | 1.32 | 1.40 |
| 314 | ♂ jun. | ......do ............... | June 11 | ....do .... | 7.18 | 4.68 | 1.30 | 1.35 |
| 315 | ♂ jun. | ......do ............... | June 11 | ....do .... | 7.33 | 4.49 | 1.58 | 1.30 |
| 316 | ♀ ad. | ......do ............... | June 11 | ....do .... | 7.60 | 4.65 | 1.25 | 1.35 |
| 317 | ♂ jun. | ......do ............... | June 11 | ....do .... | 7.70 | 4.70 | 1.60 | 1.33 |
| 343 | ♀ jun. | ......do ............... | June 14 | ....do .... | 7.10 | 4.52 | 1.28 | 1.40 |
| 344 | ♀ jun. | ......do ............... | June 14 | ....do .... | 7.20 | 4.50 | 1.24 | 1.43 |

60. **Gymnokitta cyanocephala** (Maxim.)—Maximilian's Jay.

This curious jay seems to be as eminently gregarious during the summer-months as later in the fall and winter. I frequently saw them flying from place to place in search for food, always keeping up their harsh, querulous notes, which, though somewhat jay-like, are yet peculiar to this bird. They seem to shun the dense pine-forests, and keep in the open, hilly country, where they always are found among the piñons and cedars.

| No. | Sex. | Locality. | Date. | Collector. | Wing. | Tail. | Bill. | Tarsus. |
|-----|------|-----------|-------|------------|-------|-------|-------|---------|
| 147 | ♀ ad. | Garland, Col.......... | May 26 | Henshaw. | 5.63 | 4.33 | 1.25 | 1.35 |

61. **Pica melanoleuca** V., var. **hudsonica**, Sab.—Magpie.

On the Huerfano River, May 22, this species was very common, and many of their nests were seen among the thick branches of the small trees, usually about twenty feet from the ground. These are clumsy, dome-like structures, made of coarse sticks, the bottom of the nest being lined with mud. The birds enter through a small hole left in the side, which is scarcely to be seen from the ground. One nest contained seven nearly fledged young, and as I was climbing up to examine the structure, alarmed, they clambered out, and after clinging to the sides of the nest till I had nearly reached them, they one after another launched themselves out, and soon tumbled to the ground. Meantime the parent birds made their appearance, and their cries of rage soon brought at least a dozen birds to their assistance. The whole colony kept flying around my head, screaming and scolding, and exhibiting the utmost rage; nor did they cease their outcries and efforts to distract my attention, till they had seen me well away from the neighborhood.

62. **Cyanura stelleri** (Gm.), var. **macrolopha** Bd.—Long-crested Jay.

This beautiful bird is very abundant throughout the pine-region. It is usually quite shy and suspicious, but on the Rio Grande they daily visited our camp in considerable numbers, contending with the Clarke's Crow for the fragments of food thrown away. They were, however, always on the alert, and on the first show of hostility were off to the high pines.

| No. | Sex. | Locality. | Date. | Collector. | Wing. | Tail. | Bill. | Tarsus. |
|---|---|---|---|---|---|---|---|---|
| 33 | ♂ ad. | Georgetown, Col ...... | June — | Rothrock. | 6. 05 | 5. 72 | 1. 14 | 1.71 |
| 33a | ♂ ad. | N. Fork of Platte, Col . | July — | ....do .... | 5. 53 | 5. 25 | 1. 04 | 1.54 |
| 200 | ♂ | Near Garland, Col .... | May 31 | Henshaw. | 5. 62 | 5. 54 | 1. 13 | 1.65 |
| 201 | ♂ ad. | ...... do ............. | June 5 | ....do .... | 5. 86 | 5. 58 | 1. 17 | 1.65 |
| 308 | ♂ | ...... do ............. | June 10 | ....do .... | 5. 54 | 5. 20 | 1. 05 | 1.65 |

63. *Perisoreus canadensis* (L.), var. *capitalis*, Baird.—Rocky Mountain Gray Jay.

Common in the pine-region near Garland, and also in South Park, where specimens were obtained by Dr. Rothrock. I found old birds feeding their fully fledged young the middle of June. These quite likely were second broods. The habits of this bird seem to correspond closely with those of its eastern ally, *canadensis*. It is very tame, and seems to have no feeling regarding man other than curiosity. It has a great variety of notes, and one which I often heard is a perfect imitation of the Red-tailed Hawk.

| No. | Sex. | Locality. | Date. | Collector. | Wing. | Tail. | Bill. | Tarsus. |
|---|---|---|---|---|---|---|---|---|
| 40 | ad. | Snake River, Col ..... | ......... | Rothrock. | 6. 09 | 5. 73 | 0. 85 | 1. 43 |
| 40a | | ...... do .............. | ......... | ....do .... | 5. 85 | 5. 58 | 0. 93 | 1. 39 |
| 65 | ♀ jun. | South Park, Col ...... | June 27 | ....do .... | 5. 60 | 5. 65 | 0. 88 | 1. 28 |
| 65a | Jun. | ...... do ............. | June 27 | ....do .... | 5. 90 | 5. 04 | 0. 87 | 1. 43 |
| 202 | ♂ ad. | Garland, Col.... .... | May 30 | Henshaw. | 6. 10 | 5. 93 | 0. 83 | 1. 36 |
| 223 | ♂ ad. | ...... do ............. | June 3 | ....do,.... | 5. 95 | 6. 06 | 0. 90 | 1. 37 |
| 387 | ♂ jun. | ...... do ............. | June 20 | ....do .... | 6. 04 | 5. 97 | 0. 83 | 1. 34 |

· TYRANNIDÆ (the Tyrant Flycatchers).

64. *Sayornis sayus* (Bon.)—Say's Pewee.

Not uncommon in this region. Its manner of nesting, habits, and the general character of the notes much resemble those of the eastern pewee (*S. fuscus*). A nest found June 27 beneath the eves of one of the outbuildings of the post was composed of bits of twine, shreds of cloth, and other like substances, cemented together with mud. The cavity was quite shallow, and lined thickly with horse-hair and sheep's wool. Eggs, four in number, pale yellowish-white, without spots.

65. *Contopus borealis* (Sw.)—Olive-sided Flycatcher.

A common and highly characteristic bird of the pine-region, ranging from about 7,000 feet up to timber-line. Its favorite perching-places are the tops of the high pine-stubs. From these stations it makes frequent sallies after passing insects, and seems rarely to miss its prey. When thus engaged, the clicking noise of its bill may be heard quite a distance off. About the 1st of June they had all mated, and each pair maintained a most jealous watch over the neighborhood chosen as its summer-residence, never allowing the intrusion of the larger birds to pass unnoticed. The loud call-notes of the male are at this season almost incessantly repeated. After watching the actions of several pairs, I felt sure that certain thick, tall fir-trees had been selected as the sites of their nests, but these I was not able to detect, and I do not think that the nest is finished and the eggs deposited much, if any, before the latter part of June.

Bill black; lower mandible light-brown; legs and feet black.

| No. | Sex. | Locality. | Date. | Collector. | Wing. | Tail. | Bill. | Tarsus. |
|-----|------|-----------|-------|-----------|-------|-------|-------|---------|
| 140 | ♂ ad. | Near Garland, Col .... | May 26 | Henshaw. | 4.18 | 3.05 | 0.75 | 0.54 |
| 149 | ...... | ......do ............... | May 26 | ....do .... | 4.17 | 3.00 | 0.76 | 0.60 |
| 273 | ♂ ad. | ......do .... | June 6 | ....do .... | 4.49 | 3.15 | 0.72 | 0.57 |
| 274 | ♂ ad. | ......do ............... | June 6 | ....do .... | 4.38 | 3.15 | 0.73 | 0.58 |
| 285 | ♂ ad. | ......do ............... | June 6 | ....do .... | 4.30 | 3.00 | 0.77 | 0.60 |
| 323 | ♀ ad. | Rio Grande, Col ...... | June 12 | ....do .... | 4.04 | 3.08 | 0.71 | 0.58 |
| 340 | ♀ ad. | ......do ............... | June 14 | ....do .... | 4.00 | 2.98 | 0.78 | 0.57 |

66. *Contopus virens* (L.), var. *richardsonii*, Sw.—Western Wood-Pewee.

The most abundant representative of the family. Inhabits the dark recesses of the pine-woods as well as the edges of clearings and ravines. Unlike the preceding species, which stations itself on the loftiest stubs, this flycatcher pursues its prey among the lower branches of the trees, and often descends almost to the ground to snap up a fly or moth. Its song bears a slight resemblance to that of the eastern pewee (*C. virens*), but is shorter and much more emphatic. The call-note is entirely different. A nest kindly presented by Mr. Aiken, and found by him near Fountain, Col., shows but little difference in style and structure when compared with eastern examples. It is composed mostly of sheep's wool, externally covered with bits of bark and leaves, and lined with fine grasses. Its depth, of an inch and a half, is greater than in any I have ever seen in the East, but possibly this may have been rendered necessary for the preservation of the eggs, on account of the prevalence of high winds in this locality.

| No. | Sex. | Locality. | Date. | Collector. | Wing. | Tail. | Bill. | Tarsus. |
|-----|------|-----------|-------|-----------|-------|-------|-------|---------|
| 263 | ♂ ad. | Garland, Col ......... | June 5 | Henshaw. | 2.66 | 2.95 | 0.54 | 0.51 |
| 275 | ♂ ad. | ......do ............... | June 6 | ....do .... | 2.38 | 2.68 | 0.53 | 0.54 |
| 276 | ♂ ad. | ......do ............... | June 6 | ....do .... | 2.66 | 2.95 | 0.54 | 0.51 |
| 318 | ♀ ad. | Rio Grande, Col ...... | June 17 | ....do .... | 3.25 | 2.66 | 0.64 | 0.51 |

67. *Empidonax pusillus* (Sw.)—Little Flycatcher.

Wherever willows are found growing in small clumps or fringing the streams, this flycatcher is almost certain to be found common, and it is rarely seen in the summer in other situations. Its habits and notes appear to be identical with those of its eastern analogue, from which it differs mainly in its paler coloration. Its nest is placed in the upright fork of a bush or sapling but a few feet from the ground, and is composed of grasses and fibrous material, rather loosely woven together, and lined with fine grasses. Its general appearance is much like that of the nest of the Yellow Warbler (*D. æstiva*), but it is not nearly so compact nor artistic.

| No. | Sex. | Collector. | Date. | Collector. | Wing. | Tail. | Bill. | Tarsus. |
|---|---|---|---|---|---|---|---|---|
| 131 | ♂ ad. | Garland, Col.......... | May 25 | Henshaw | 2.81 | 2.64 | 0.55 | 0.67 |
| 164 | ♂ ad. | ......do .............. | May 27 | ....do .... | 2.83 | 2.65 | 0.52 | 0.68 |
| 165 | ♂ ad. | ......do .............. | May 28 | ....do .... | 2.75 | 2.49 | 0.55 | 0.68 |
| 166 | ♂ ad. | ......do .............. | May 28 | ....do .... | 2.82 | 2.53 | 0.58 | 0.68 |
| 174 | ♂ ad. | ......do .............. | May 29 | ....do .... | 3.00 | 2.60 | 0.51 | 0.65 |

68. *Empidonax obscurus* (Sw.)—Wright's Flycatcher.

Apparently a rather rare summer-resident in this region. But two specimens were taken; these in worn breeding-plumage in June.

69. *Empidonax hammondi*, Vesey.—Hammond's Flycatcher.

While collecting in the mountains near the Rio Grande during the middle of June, I saw several pairs of this little Flycatcher. I found them in the pine-region on the small streams fringed here and there with alders, but they seemed to hunt by preference among the contiguous pines. Their habits are somewhat peculiar, and have but little of the dash and spirit which characterize most of the birds of this family, and especially the Least Flycatcher (*E. minimus*), with which this species seems most nearly related. After snapping up a passing insect, it resumes its perch upon some low limb, and remains nearly motionless for a time, giving an occasional listless jerk of the tail. The notes are very feeble, the most so of any flycatcher I am acquainted with, and consist of a soft pit, varied with a low, lisping whistle.

| No. | Sex. | Locality. | Date. | Collector. | Wing. | Tail. | Bill. | Tarsus. |
|---|---|---|---|---|---|---|---|---|
| 341 | ♀ ad. | Rio Grande, Col ...... | June 14 | Henshaw. | 2.64 | 2.13 | 0.41 | 0.61 |

70. *Empidonax flaviventris* Bd., var. *difficilis*, Bd.—Western Yellow bellied Flycatcher.

An occasional pair seen in the same locality as the preceding, where it was found inhabiting the deep shady glens of the pine-woods, often near a running stream. It is a rather energetic insect-hunter, continually swooping down after passing insects, and when waiting for its prey moving its tail with nervous and excited jerks.

| No. | Sex. | Locality. | Date. | Collector. | Wing. | Tail. | Bill. | Tarsus. |
|---|---|---|---|---|---|---|---|---|
| 342 | ♀ ad. | Rio Grande, Col ...... | June 14 | Henshaw. | 2.57 | 2.37 | 0.43 | 0.62 |

ALCEDINIDÆ (the Kingfishers).

71. *Ceryle alcyon* (L.)—Belted Kingfisher.

An occasional resident on the creeks, and not uncommon on the Rio Grande.

CAPRIMULGIDÆ (the Goatsuckers).

72. *Antrostomus nuttalli* (Aud.)—Nuttall's Whippoorwill.

This bird is found in varying numbers throughout this entire region, but everywhere it is much more numerous than its cousin of the Eastern States. It makes its appearance in the deeply-shadowed portions of the river-bottoms a few minutes before dusk, and, as soon as night settles down, the rather mournful note of *poor-will, poor-will*, may be heard coming from the edges of the woods, and even from the sage-brush plains. Their notes are most often noticed in early evening, and again just before dawn, but not infrequently their song is heard through the entire night. When on the wing after insects, their flight consists of rapid, irregular turnings and windings, which are prolonged but a moment or so, when they alight, often on a fallen log, but usually on the bare ground. Occasionally, at dusk, I have seen them alight almost at my feet, without betraying any sense of my presence. When flying they emit a constantly-repeated clucking note, which is, I think, common to both sexes. Their eggs are pure white, without spots, and are deposited on the ground during the latter part of June.

73. *Chordeiles popetue* (Vieill.), var. *henryi*, Cass.—Western Night-Hawk.

An exceedingly abundant species everywhere in the vicinity of water. Often seen at noonday flying over the surface of stagnant pools, catching the insects which swarm in such places. They are, however, usually most active just before dusk, and on the banks of the Rio Grande I have seen them at this time make their appearance in hundreds.

TROCHILIDÆ (the Humming-Birds).

74. *Selasphorus platycercus* (Sw.)—Broad-tailed Humming-Bird.

This, the only humming-bird seen in this region, was found in very great numbers. Though most common on the creeks, at an altitude of about 7,000 feet, it also reaches well up timber-line. A nest, found June 14, was saddled to a horizontal limb of a small spruce; a second, taken the 19th, was built on a small, swinging branch of a cottonwood. They are less artistic structures than usual with birds of this family, and are composed of cottony substances from plants, covered externally with bits of bark and moss. Both contained two white eggs, perfectly fresh. During the mating, and perhaps also through the entire breeding season, the flight of the male is always accompanied by a curious, loud, metallic, rattling noise, which he is enabled to produce in some way by means of the attenuation of the outer primaries. This is, I think, intentionally made, and is analogous to the love-notes of other birds. Though I saw many of these birds in the fall, it was only very rarely that this whistling noise was heard, and then with greatly diminished force.

| No. | Sex. | Locality. | Date. | Collector. | Wing. | Tail. | Bill. |
|---|---|---|---|---|---|---|---|
| 126 | ♂ ad. | Garland, Col .......... | May 25 | Henshaw... | 1.97 | 1.44 | 0.44 |
| 127 | ♂ ad. | ......do .............. | May 25 | ....do ...... | 1.98 | 1.45 | 0.65 |
| 128 | ♂ ad. | ......do .............. | May 25 | ....do ...... | 1.90 | 1.50 | 0.65 |
| 128a | ♂ ad. | ......do .............. | May 25 | ....do ...... | 1.97 | 1.44 | 0.65 |
| 128b | ♂ ad. | ......do .............. | May 27 | ....do ...... | 1.93 | 1.40 | 0.68 |
| 157 | ♂ ad. | ......do .............. | May 28 | ....do ...... | 1.88 | 1.50 | 0.72 |
| 159 | ♂ ad. | ......do .............. | May 28 | ....do ...... | 1.90 | 1.36 | 0.64 |
| 347 | ♀ ad. | ......do .............. | May 14 | ....do ...... | 1.96 | 1.48 | 0.72 |

PICIDÆ (the Woodpeckers).

**75. *Picus villosus*, L., var. *harrisii*, Aud.—Harris's Woodpecker; Western Hairy Woodpecker.**

An inhabitant of the pine-woods, where it is found abundant up to timber-line.

| No. | Sex. | Locality. | Date. | Collector. | Wing. | Tail. | Bill. | Tarsus. |
|---|---|---|---|---|---|---|---|---|
| 228 | ♂ ad. | Mountains near Garland, Col. | June 4 | Henshaw. | 3.98 | 2.83 | 0.67 | 0.65 |

**76. *Picus pubescens* L., var. *gairdneri* Aud.—Gairdner's Woodpecker; Western Downy Woodpecker.**

While the preceding is perhaps the most characteristic woodpecker of the West, the present species is quite common. During my stay in this vicinity three were secured; a pair in a grove of cottonwoods, the third in the pines at an elevation of perhaps 10,000 feet. These were all breeding.

| No. | Sex. | Locality. | Date. | Collector. | Wing. | Tail. | Bill. | Tarsus. |
|---|---|---|---|---|---|---|---|---|
| 146 | ♀ ad. | Near Garland, Col .... | May 26 | Henshaw. | 4.10 | 2.84 | 0.70 | 0.65 |
| 222 | ♀ ad. | ......do ..... | June 3 | ....do .... | 4.09 | 3.00 | 0.67 | 0.63 |
| 238 | ♂ ad. | ......do ..... | June 4 | ....do .... | 3.95 | 2.81 | 0.67 | 0.60 |

**77. *Picoides arcticus* (Sw.), var. *dorsalis*, Bd.—Striped-backed Woodpecker.**

A single individual was taken in June at an altitude of 10,000 feet. It was a female, and from the swollen condition of the abdomen was evidently incubating.

| No. | Sex. | Locality. | Date. | Collector. | Wing. | Tail. | Bill. | Tarsus. |
|---|---|---|---|---|---|---|---|---|
| 221 | ♀ ad. | Mountains near Garland, Col | June 3 | Henshaw. | 4.72 | 3.43 | 1.05 | 0.77 |
| 70 | ♀ ad. | South Park, Col ...... | June 28 | Rothrock. | 4.78 | 3.46 | 1.05 | 0.77 |

**78. *Sphyrapicus varius* (L.), var. *nuchalis*, Bd.—Red-naped Woodpecker.**

A very common inhabitant of the cottonwoods; rarely seen among the pines. Found by Dr. Rothrock in South Park, among the aspens, up to 12,000 feet. The young, in nesting-plumage, were taken July 12.

| No. | Sex. | Locality. | Date. | Collector. | Wing. | Tail. | Bill. | Tarsus. |
|---|---|---|---|---|---|---|---|---|
| 136 | ♀ ad. | Near Garland, Col .... | May 25 | Henshaw. | 5.04 | 3.42 | 0.85 | 0.77 |
| 145 | ♂ ad. | ......do ..... | May 26 | ....do .... | 4.98 | 3.10 | 0.86 | 0.77 |
| 153 | ♂ ad. | ......do ..... | May 27 | ....do .... | 5.07 | 3.40 | 0.90 | 0.82 |
| 154 | ♀ ad. | ......do ..... | May 27 | ....do .... | 4.98 | 3.15 | 0.88 | 0.73 |
| 172 | ♀ ad. | ......do ..... | May 29 | ....do .... | 4.88 | 3.35 | 0.88 | 0.77 |
| 226 | ♂ ad. | ......do ..... | June 2 | ....do' .... | 4.92 | 2.90 | 0.88 | 0.77 |
| 228 | ♀ ad. | ......do ..... | June 2 | ....do .... | 4.80 | 3.15 | 0.90 | 0.77 |
| 236 | ♂ ad. | ......do ..... | June 4 | ....do .... | 4.88 | 2.91 | 0.93 | 0.80 |
| 267 | ♂ ad. | ......do ..... | June 5 | ....do .... | 4.80 | 3.15 | 0.90 | 0.77 |

79. *Sphyrapicus thryroideus* (Cass.)—Black-breasted Woodpecker.

This species was first made known to science through a description by Cassin, published in December, 1851, in Pr. A. N. Sc. In 1857, Dr. Newberry published a description of Williamson's Woodpecker (*S. williamsonii*) from specimens obtained by Lieutenant Williamson's expedition, since which time the two species have been accepted by ornithologists as perfectly valid, the true relationship of the two being wholly unsuspected. While near Garland, I obtained abundant proof of the specific identity of the two birds in question, *williamsonii* being the male of *thryroidens*. Though led to suspect this, from finding the two birds in suspicious proximity, it was some time before I could procure a pair actually mated. A nest was at length discovered, excavated in the trunk of a live aspen, and both the parent birds were secured as they flew from the hole, having just entered with food for the newly-hatched young. As regards the sexual differences of coloration, the case of *thryroidens* is wholly unique. In this species, the colors of the female are radically different from those of the male. With this single exception, as far as known, the differences of color between the sexes in the family of woodpeckers are confined mainly to the absence or less amount of the bright-crimson or red patches about the head. The species is a resident of the pine-woods, abundant at an altitude of 10,000 feet, and doubtless is found at least up to the pine-limit. Except in evincing at all times a marked preference for pine-timber, rarely indeed alighting on any of the deciduous trees, their habits and notes seem to correspond pretty closely with those of *Sphyropicus nuchalis*. The stomachs of all the specimens examined contained nothing but insects and larvæ. As, however, the structure of the tongue is identical with *varius*, the species may possibly, in winter and spring, when other food is scarce, feed upon the inner bark of the deciduous trees, as the common sapsuckers (*varius* and varieties) are well known to do. I never noticed anything, however, which would lead me to suppose this. The nest mentioned above was dug to the depth of seven inches, and was one and three-fourths inches in diameter. The egg-shells had not been removed; and one which is tolerably whole shows their similarity with those of *varius*, but appears a trifle larger.

| No. | Sex. | Locality. | Date. | Collector. | Wing. | Tail. | Bill. | Tarsus. |
|-----|------|-----------|-------|------------|-------|-------|-------|---------|
| 198 | ♂ ad. | Mountains near Garland, Col. | May 30 | Henshaw. | 5.40 | 3.75 | 0.95 | 0.85 |
| 217 | ♀ ad. | ......do ....:........ | June 2 | ....do .... | 5.08 | 3.28 | 0.91 | 0.77 |
| 219 | ♂ ad. | ......do ............ | June 3 | ....do .... | 5.20 | 3.75 | 0.90 | 0.85 |
| 220 | ♀ ad. | ......do ............ | June 3 | ....do .... | 5.37 | 3.80 | 0.95 | 0.85 |
| 234 | ♂ ad. | ......do ............ | June 4 | ....do .... | 5.28 | 3.85 | 1.07 | 0.84 |
| 235 | ♂ ad. | ......do ............ | June 4 | ....do .... | 5.25 | 3.70 | 0.95 | 0.87 |
| 329 | ♀ ad. | ......do ............ | June 12 | ....do .... | 5.50 | 3.63 | 0.95 | 0.81 |
| 334 | ♂ ad. | ......do ............ | June 13 | ....do .... | 5.45 | 3.63 | 1.01 | 0.80 |
| 335 | ♀ ad. | ......do ............ | June 13 | ....do .... | 5.30 | 3.60 | 0.93 | 0.80 |

80. *Melanerpes erythrocephalus* (L.)—Red-headed Woodpecker.

Dr. Rothrock saw this species in South Park, in July, at an elevation of nearly 10,000 feet, and obtained a single specimen at the Twin Lakes at about the same height. Not seen near Garland, but noted on the Huerfano River, eighty miles northeast of this post, from which I judge it occurs generally, but sparingly, in Colorado.

81. *Colaptes auratus* (L.), var. *mexicanus*, Sw.—Red-shafted Woodpecker.

Abundant everywhere, frequenting indifferently the deciduous and coniferous trees up to timber-line. A natural cavity in a cottonwood-tree contained three fresh eggs. This was May 22. A male, taken at Fort Garland, is noticeable as having distinct black markings in the red maxillary patch.

| No. | Sex. | Locality. | Date. | Collector. | Wing. | Tail. | Bill. | Tarsus. |
|---|---|---|---|---|---|---|---|---|
| 189 | ♀ ad. | Garland, Col........... | May 29 | Henshaw. | 6.47 | 5.05 | 1.50 | 1.10 |
| 272 | ♂ ad. | ......do ............... | June 6 | ....do .... | 6.50 | 4.84 | 1.35 | 1.08 |
| 284 | ♀ ad. | ......do ............... | June 7 | ....do .... | 6.23 | 4.30 | 1.42 | 1.08 |

STRIGIDÆ (the Owls).

82. *Bubo virginianus* (Gm.), var. *arcticus* Sw.—Western Great-horned Owl.

But one seen among the mountains. Said to be not uncommon in fall and winter, descending to the plains.

FALCONIDÆ (the Falcons).

83. *Falco sparverius*, L.—Sparrow-Hawk.·

Common everywhere below 10,000 feet.

| No. | Sex. | Locality. | Date. | Collector. | Wing. | Tail. | Bill. | Tarsus. |
|---|---|---|---|---|---|---|---|---|
| 184 | ♀ ad. | Garland, Col.......... | May 28 | Henshaw.. | 7.77 | 5.38 | 0.47 | 1.50 |
| 4 | ♂ ad. | Twin Lakes, Col ...... | Aug. — | Rothrock.. | 7.20 | 5.13 | 0.45 | 1.42 |

84. *Buteo borealis* (Gm.), var. *calurus*, Cass.—Western Red-tail.

Not common. A fine adult bird was noticed on several occasions, soaring far above a tract of pine-woods.

85. *Archibuteo ferrugineus* (Licht.)—California Squirrel-Hawk.

Not met with during the summer. In November, this species was numerous on the plains about Pueblo, and also near Colorado City. Unquestionably breeds in the mountains.

86. *Archibuteo lagopus* (Brünn.), var. *sancti-johannis*, Gm.

Not seen near Garland. Probably, however, spends the summer in the mountains, as the species was a common one on the plains near Colorado City.

CATHARTIDÆ (the American Vultures).

87. *Rhinogrypus aura* (L.)—Red-headed Vulture.

Apparently rather rare in this region. A few seen during the month of June, sailing high in air.

COLUMBIDÆ (the Pigeons).

88. *Zenaidura carolinensis* (L.)—Carolina Dove.

Abundant between 7,000 and 10,000 feet.

TETRAONIDÆ (the Grouse).

89. *Canace obscura* (Say).—Dusky Grouse.

Abundant. Found during the summer on the mountain-ridges, in groves of pine and aspen, from 7,000 feet up to timber-line. Dr. Rothrock obtained specimens and found the species numerous at an altitude of from 10,000 to 12,000 feet; those at the former elevation frequenting the cottonwood-groves, while at the latter they were found in the pines only. It is quite tame and unsuspicious, and when forced to fly, which it does unwillingly, takes to the nearest tree, and then, as if incapable of further effort, stands gazing at the intruder with outstretched neck till brought down by a shot from a gun or a revolver. A nest found June 16 contained seven eggs just on the point of hatching. The nesting-site was a peculiar one, being in an open glade, where the grass had been recently burned off. The nest proper was a slight collection of dried grass, placed in a depression between two tussocks, there apparently having been no attempt made at concealment. The eggs are pale yellowish-white, spotted irregularly with reddish brown; length, 1.95; diameter, 1.39.

| No. | Sex. | Locality. | Date. | Collector. | Wing. | Tail. | Bill. | Tarsus. |
|-----|------|-----------|-------|------------|-------|-------|-------|---------|
| 129 | ...... | ........................ | .......... | ............ | 9.35 | 7.90 | 0.75 | 1.60 |
| 63 | | ........................ | .......... | ............ | 9.10 | 7.40 | 0.75 | 1.60 |
| 312 | ♂ ad. | Rio Grande, Col....... | June 12 | Henshaw. | 9.70 | 7.50 | 0.75 | 1.75 |

90. *Centrocercus urophasianus* (Bon.)—Sage-Cock.

Not found by me in vicinity of Garland. A single flock was seen by Dr. Rothrock on the headwaters of the Arkansas in August.

| No. | Sex. | Locality. | Date. | Collector. | Wing. | Tail. | Bill. | Tarsus. |
|-----|------|-----------|-------|------------|-------|-------|-------|---------|
| 7 | ♀ (?) | Twin Lakes, Col...... | Aug. — | Rothrock. | 10.20 | 7.50 | 1.23 | 1.78 |

91. *Lagopus leucurus*, Sw.—White-tailed Ptarmigan.

This beautiful species was found by Dr. Rothrock abundant in the mountains of South Park during the latter part of June and July. It ranges from the timber-line to the summits of the highest peaks, showing always a preference for rocky localities. It was found at the extreme height of 14,400 feet, in the most sterile districts, where no vegetation existed. Their habits, as observed by Dr. Rothrock, were as follows: During the heat of the day they remain quiet beneath the shelter of the rocks, but in early morning and evening were seen running over the ground, actively engaged in searching for food, and keeping up a constant chirruping. They usually seemed entirely devoid of fear, allowing themselves to be almost trodden upon before taking flight, but sometimes were very shy and wild. The young birds well grown were seen July 10, so that the eggs are deposited by the first of May. The nest is simply a small cavity scratched in the earth under a projecting rock, sometimes with a slight lining of sticks and grasses, but oftener without. In winter they descend into the timber, and are then so tame as to be often killed with clubs.

| No. | Locality. | Date. | Collector. | Wing. | Tail. | Bill. | Tarsus. |
|-----|-----------|-------|-----------|-------|-------|-------|---------|
| 101 | South Park, Col.............. | July — | Rotbrock . | 7.30 | 4.25 | 0.58 | 1.15 |
| 102 | ......do...................... | July — | .... do .... | 7.27 | 4.58 | 0.56 | 1.24 |

### CHARADRIIDÆ (the Plovers).

**92. *Ægialitis vociferus* (L.)—Kildeer-Plover.**

Frequently met with in pairs in June along the water-courses. I found the young just from the nest June 14.

**93. *Ægialitis montanus* (Towns.)—Rocky-Mountain Plover.**

This species was met with but in one locality, on the dry plains near the Rio Grande. It is to be regretted that lack of time did not allow a more careful examination of the habits of this little known species. While riding rapidly along in an ambulance I saw quite a number, and shot three as they ran from before the horses and halted a few feet from the road. From their actions I was certain that their eggs were near by, but a short search did not reveal them. Upon dissecting a female, I found an egg nearly ready to be deposited. This was June 10. They were very tame, running along the ground a few feet ahead, and uttering a low, croaking note.

| No. | Sex. | Locality. | Date. | Collector. | Wing. | Tail. | Bill. | Tarsus. |
|-----|------|-----------|-------|-----------|-------|-------|-------|---------|
| 305 | ♂ ad. | Rio Grande, Col....... | June 10 | Henshaw . | 5.93 | 2.95 | 0.81 | 1.45 |
| 306 | ♀ ad. | ......do ............. | June 10 | .... do .... | 5.80 | 2.75 | 0.89 | 1.48 |
| 307 | ♂ ad. | ......do ............. | June 10 | .... do .... | 5.50 | 2.59 | 0.85 | 1.38 |

### SCOLOPACIDÆ (the Snipes).

**94. *Tringoides macularius* (L.)—Spotted Sandpiper.**

A few individuals were seen in June. Doubtless breed.

### RECURVIROSTRIDÆ (the Stilts and Avocets).

**95. *Recurvirostra americana*, Gm.—Avocet.**

An abundant summer-resident on the shores of the alkali lakes northwest of Garland. At the time of my visit, June 21, the greater number evidently had young, as I found many broken egg-shells along the shores. Wherever I went, the parent-birds manifested the greatest solicitude, flying about my head in flocks, and uttering their loud, hoarse cries. They are adept swimmers, freely alighting on the surface of the water, where they float buoyantly and gracefully. The food of these birds, as well as the stilts and ducks, seems to consist almost exclusively, at this season, of the larvæ of single species of insect, with which the alkaline water fairly swarms. The crops of the birds examined were filled with these and a few water-beetles. They deposit their eggs in a slight hollow scratched in the sand and lined with weeds. These are four in number, of a dull olive-brown color, blotched all over with black. Length, 2.00–1.43; (No. 2,) 1.85–1.07; (No. 3,) 1.91–1.43; (No. 4,) 1.91–1.42.

| No. | Sex. | Locality. | Date. | Collector. | Wing. | Tail. | Bill. | Tarsus. |
|-----|------|-----------|-------|-----------|-------|-------|-------|---------|
| 389 | ♂ ad. | Alkali lakes, Col...... | June 21 | Henshaw. | 8.80 | 3.73 | 3.60 | 3.50 |
| 390 | ♂ ad. | ......do ............... | June 21 | .... do .... | 8.90 | 3.97 | 3.76 | 3.40 |
| 391 | ♀ ad. | ......do ............... | June 21 | .... do .... | 9.00 | 3.70 | 3.36 | 3.32 |

96. *Himantopus nigricollis*, Vieill.—Stilt.

Nearly as abundant as the preceding. Like it, the eggs had been hatched, and I found the young but just from the nest. One nest, built in the same manner as the one mentioned above, contained fresh eggs. They are indistinguishable from those of the preceding species except by their smaller size.

Length, 1.74–1.31; (No. 2,) 1.74–1.27; (No. 3,) 1.74–1.27.

| No. | Sex. | Locality. | Date. | Collector. | Wing. | Tail. | Bill. | Tarsus. |
|-----|------|-----------|-------|-----------|-------|-------|-------|---------|
| 393 | ♂ ad. | Alkali lakes, Col...... | June 21 | Henshaw. | 9.25 | 3.42 | 2.60 | 4.48 |
| 398 | ♀ ad. | ......do ............... | June 21 | .... do .... | 8.50 | 3.57 | 2.55 | 3.95 |
| 400 | ♀ ad. | ......do ............... | June 22 | .... do .... | 8.60 | 3.28 | 2.48 | 3.94 |
| 404 | ♀ ad. | ......do ............... | June 22 | .... do .... | 8.60 | 3.21 | 2.51 | 4.12 |

RALLIDÆ (the Rails).

97. *Fulica americana*, Gm.—Coot.

Very numerous at the lakes. They breed in colonies among the rushes, the nests often being but a few feet apart. They are very bulky structures, composed of weeds and rushes raised to a height of several inches from the surface of the water, so that the eggs are kept perfectly dry, and are moored to the stems of the surrounding reeds. The greatest number of eggs found in one nest was eleven, and most contained from five to seven, showing that they were not through laying. This was June 22.

ANATIDÆ (the Ducks).

The following ducks were found at this same locality. Besides those given, others occur here as summer-residents, but owing to their shyness the species could not be satisfactorily determined:

98. *Anas boschas*, L.—Mallard.

Rather numerous. Breeding.

99. *Querquedula cyanoptera* (Vieill.)—Red-breasted Teal.
100. *Querquedula discors* (L.)—Blue-winged Teal.

Both observed in considerable numbers. Several teals' nests were found with partial complements of eggs; but owing to the absence of the owners, their identity could not be determined.

**101.** *Nettion carolinensis* (Gm.)—Green-winged Teal.

Also common. A nest containing ten eggs was found under a sage-bush, perhaps thirty feet from the water's edge. A deep hollow had been scooped in the sand, and lined warmly with fine grasses and down, evidently taken from the bird's own breast, which was plucked nearly bare. The eggs are of a pale-yellowish color, and average 1.81 in length by 1.31 in diameter.

**102.** *Spatula clypeata* (L.)—Shoveler.

Rather common.

**103.** *Chaulelasmus streperus* (L.)—Gadwall.

A few seen.

PODICIPIDÆ (the Grebes).

**104.** *Podiceps auritus* (L.), var. *californicus*, Heerm.—American Eared Grebe.

Common. A colony was found breeding, their nests being placed in a bed of reeds in the middle of a small pond. The nests were slightly hollowed piles of decaying weeds and rushes, just raised above the surface of the water, upon which they floated. Each nest contained three eggs, most of them being fresh, but a few were somewhat advanced. As in every case the eggs were entirely covered by a pile of vegetable material, and as in no case the birds were found incubating, even where the eggs gave evidence of the fact, it seems highly probable that their hatching is dependent more or less upon artificial heat, which must be induced by the effect of the hot sun.

The eggs vary little in shape, are considerably elongated, one end being slightly more pointed than the other. They vary in length from 1.70 to 1.80; in breadth, 1.18 to 1.33. Color a faint yellowish white, usually much stained by contact with the nest. The texture is generally quite smooth; in others roughened by a chalky deposit.

| No. | Sex. | Locality. | Date. | Collector. | Wing. | Bill. | Tarsus. |
|-----|------|-----------|-------|------------|-------|-------|---------|
| 414 | ♂ ad. | Alkali lakes, Col...... | June 23 | Henshaw . ...... | 5.33 | 1.02 | 1.52 |

SECTION III.

Leaving Southern Colorado July 2, I arrived at Fort Wingate, N. Mex., the 12th. During a week's delay, attendant on fitting out the several parties, short trips were made into the neighboring region; Dr. Newberry, jr., and myself accompanying the party on each occasion, and making collections in natural history. Owing to the rather desolate nature of the country, these, however, were not very extensive; the birds especially being found rather scarce. Starting from Wingate July 19, a southwesterly course was pursued, our destination being Apache, Ariz., where we arrived August 2. During this interval, as we moved slowly, I was enabled to spend considerable time in making collections, and some very interesting results were obtained, especially in the way of birds. From August 2 till September 6, collections were made by Dr. Newberry and myself in the vicinity of Apache and the adjoining White Mountains. This region proved very interesting ornithologically, and, indeed, the general collections made here were perhaps larger than during any other equal period through the season. From here southward, quite a distinct change in the character of the avifauna was noticed, and a number of species were noted either in the vicinity of Apache or a few miles to the northward, that probably find their northward limit here. Such are *Pyranga hepatica*, *Peucœa ruficeps*, var. *boucardi*, *Cyanocitta ultramarina*, var. *arizonœ*, *Setophaga picta*, *Melanerpes formicivorus*, &c. The Gila River was crossed at a point some sixty miles south of Apache, and a few days' stay along the river gave valuable results in natural history. We arrived at Fort Bowie, the southernmost point reached, October 6. From here our route led northward to the Gila River, which was followed to its sources in New Mexico, after which a general northward course was taken for Wingate, which was reached November 27, when the field-work ended. Though no new species were detected, one (*Eugenes fulgens*) was added to our fauna, and numerous specimens secured of rare and little-known species, while the geographical range of quite a number was widely extended. During the last month, the results in natural history, owing to the lateness of the season, were rather meager. Thus, the region in which most of the observations following were made may be stated in general terms to be the southeastern portion of Arizona and Southwestern New Mexico.

## TURDIDÆ (the Thrushes).

1. *Turdus migratorius*, L.—Robin.

Is scarcely to be regarded as a common bird, but it was frequently met with in the timber along the streams and in the mountains. At the Old Crater, forty miles south of Zuni, N. Mex., the species was present in large flocks the 2d of November. The surrounding hills are covered with low scrubby cedars, and upon the berries this and other species largely subsist at this late season. It doubtless winters in this region.

| No. | Sex. | Locality. | Date. | Collector. | Wing. | Tail. | Bill. | Tarsus. |
|-----|------|-----------|-------|-----------|-------|-------|-------|---------|
| 27 | ♀ ad. | Wingate, N. Mex...... | July 12 | Newberry. | 5.23 | 4.10 | 0.73 | . 1.15 |

2. *Turdus pallasi* Cab., var. *nanus*, Aud.—Dwarf Hermit-Thrush.

This variety of the hermit-thrush was met with October 19, along the small streams in the mountain-cañons, near Fort Bowie, Southeastern Arizona, and along the Gila River to its sources in New Mexico, where I found it as late as November 8. It is undoubtedly abundant during the fall-migration throughout a very large extent of country, as, wherever seen, it was in large numbers. Its habits seem to differ in no noteworthy respect from the allied forms. It appears fond of solitude, and prefers the thickest and shadiest thickets, where it is constantly busied in searching among the leaves for seeds and insects. Its small size is apparent at first sight, and serves even when alive to distinguish it from either var. *auduboni* or *pallasi*.

| No. | Sex. | Locality. | Date. | Collector. | Fresh. | W. | T. | B. | Tar. |
|---|---|---|---|---|---|---|---|---|---|
| 942 | ♂ | Near Bowie, Ariz .. | Oct. 19 | Henshaw. | ..... | ..... | 3.45 | 2.81 | 0.45 | 1.08 |
| 943 | ♀ | ......do ............ | Oct. 19 | ....do .... | 6.25 | 10.00 | 3.37 | 2.74 | 0.45 | 1.12 |
| 944 | ♀ | ......do ............ | Oct. 19 | ....do .... | 6.12 | 9.74 | 3.25 | 2.55 | 0.49 | 1.04 |
| 945 | ♀ | ......do ............ | Oct. 19 | ....do .... | 6.25 | 10.18 | 3.49 | 2.74 | 0.42 | 1.10 |
| 982 | ♂ | ......do ............ | Nov. 5 | ....do .... | 6.55 | 10.87 | 3.61 | 2.74 | 0.49 | 1.14 |

3. *Oreoscoptes montanus* (Towns.)—Sage-Thrasher; Mountain-Mocker.

Quite numerous in the vicinity of Fort Wingate, N. Mex., where as elsewhere its favorite abode was the sage-brush plains. Here Dr. Newberry, jr., found a nest July 14, containing eggs just ready to be hatched. The nest was as usual a large bulky structure of sticks and twigs, lined with fine rootlets, and placed in the top of a sage-bush. The species was met with at various localities along our route. In the fall it is not unusual to find small companies of from five to ten associating together. They are nearly always shy and suspicious of the presence of man.

| No. | Sex. | Locality. | Date. | Collector. | Wing. | Tail. | Bill. | Tarsus. |
|---|---|---|---|---|---|---|---|---|
| 513 | ♀ ad. | Cave Spring, Ariz..... | Aug. 1 | Henshaw . | 3.86 | 3.80 | 0.68 | 1.17 |
| 712 | ♂ ad. | Apache, Ariz ......... | Sept. 7 | ....do .... | 3.92 | 4.02 | 0.65 | 1.15 |
| 780 | ♀ jun. | Gila River, Ariz....... | Sept.15 | McGee ... | 3.92 | 3.90 | 0.66 | 1.17 |
| 601 | ♀ jun. | Goodwin, Ariz......... | Sept.19 | Henshaw . | 3.90 | 3.66 | 6.62 | 1.13 |
| 126 | Jun. | Bowie, Ariz........... | Oct. 10 | Newberry. | 3.77 | 3.89 | 0.65 | 1.17 |
| 14 | ♀ jun. | ......do ............. | Oct. 7 | ....do .... | 3.75 | 3.65 | 0.75 | 1.19 |

4. *Harporynchus crissalis* Henry.—Red-vented Thrasher.

This was by no means an uncommon species in the cañons at the base of Mount Turnbull, eight miles west of old Fort Goodwin, Ariz. It frequented the brush along the cañon-sides, and it was only after much trouble that I succeeded in obtaining a single specimen. A second's glimpse, as it darted far ahead from some low bush into the thick brush, was usually the only proof to be had of its presence. I judge it to be generally, but sparingly, distributed in this part of Arizona, and perhaps the southeastern part of New Mexico, as on several occasions in the mezquite-covered plains along the Gila River, I saw a few curve-billed thrushes, which were most likely of this species.

Iris brown; bill black; feet plumbeous-brown.

| No. | Sex. | Locality. | Date. | Collector. | Wing. | Tail. | Bill. | Tarsus. |
|---|---|---|---|---|---|---|---|---|
| 799 | ♂ jun. | Mt. Turnbull, Ariz .... | Sept. 19 | Henshaw. | 3.92 | 5.75 | 1.14 | 1.32 |

5. *Mimus polyglottus* (L.)—Mocking-bird. ·

I procured a young bird of this species at Inscription Rock, N. Mex., from among half a dozen others, and a few were afterward seen on the road to Apache, Ariz. They were very shy and restless. Said to be a common summer-resident of Arizona. (Coues.)

7 o s

CINCLIDÆ (the Dippers).

6. *Cinclus mexicanus,* Sw.—Water-Ouzel.

Found among the rapids of the streams in the White Mountains, Arizona, where, however, it did not appear to be numerous. A pair of these interesting birds were seen on a small isolated pond in the high pine-woods, where they seemed as perfectly at home as in their customary haunts on the most turbulent streams. Also met with on a mountain-stream near Tulerosa, N. Mex.

SAXICOLIDÆ (the Saxicolas).

7. *Sialia mexicana,* Sw.—Western Bluebird.

This species was not observed till July 23 at Inscription Rock. This appeared to be a favorite locality, and large numbers of both old and young were congregated together in the piñon and cedar trees. Their habits at this season do not differ notably from the other species. From here southward they were frequently seen, commonly among the pines. At Apache in August I found them in large flocks in the pine-woods, and accompanied by flocks of warblers, nuthatches, and titmice, to which they seemed to act as leaders, the whole flock following their flight from tree to tree.

| No. | Sex. | Locality. | Date. | Collector. | Wing. | Tail. | Bill. | Tarsus. |
|-----|------|-----------|-------|------------|-------|-------|-------|---------|
| 464 | ♀ jun. | Inscription Rock, N. Mex. | July 23 | Henshaw. | 4.30 | 2.71 | 0.40 | 0.83 |
| 465 | ♂ | ......do ............... | July 23 | .... do .... | 4.07 | 2.70 | 0.45 | 0.79 |
| 527 | ♂ | Apache, Ariz........... | Aug. 26 | .... do .... | 4.10 | 2.85 | 0.43 | 0.68 |
| 679 | ♂ ad. | ......do ............... | Sept. 1 | .... do .... | 4.30 | 2.95 | 0.42 | 0.83 |
| 975 | ♂ ad. | Gila River, Southwestern New Mexico. | Oct. 28 | .... do .... | 4.45 | 2.98 | 0.47 | 0.60 |
| 962 | ♂ ad. | ......do ............... | Oct. 28 | .... do .... | 4.43 | 2.98 | 0.50 | 0.85 |

8. *Sialia arctica,* Sw.—Rocky-Mountain Bluebird.

On leaving the mountainous region of Southern Colorado, this species was apparently left behind, nor did I again see it till the middle of November, when I found very large flocks in the neighborhood of a spring at the salt lake south of Zuni, N. Mex. They doubtless winter here.

| No. | Sex. | Locality. | Date. | Collector. | Wing. | Tail. | Bill. | Tarsus. |
|-----|------|-----------|-------|------------|-------|-------|-------|---------|
| 998 | ♂ ad. | Salt Lake, N. Mex .. | Nov. 19 | Henshaw. | 4.58 | 3.03 | 0.50 | 0.85 |
| 1006 | ♂ ad. | ......do ............ | Nov. 19 | ..., do .... | 4.75 | 2.27 | 0.50 | 0.93 |

SYLVIIDÆ (the Sylvias).

9. *Regulus calendula* (L.)—Ruby-crowned Wren.

Common in the White Mountains of Arizona in August, where doubtless it also breeds. Also seen at Camp Grant September 24, and common on the Gila River, among the cottonwoods, October 19.

10. *Polioptila cœrulea* (L.)—Blue-gray Flycatcher.

Met with first a little south of Apache, Ariz., and an occasional individual seen afterward, usually among the oak-trees. Their habits at this season are much like the warblers. They were very active, passing rapidly in and out among the small branches, and seeming in such hurry as to scarcely begin their examination of one tree ere they were off to another. Now and then I saw one catching insects on the wing.

| No. | Sex. | Locality. | Date. | Collector. | Wing. | Tail. | Bill. | Tarsus. |
|---|---|---|---|---|---|---|---|---|
| 720 | ♀ | South of Apache, Ariz. | Sept. 8 | Henshaw. | 1.91 | 2.24 | 0.33 | 0.65 |
| 898 | ♂ | San Pedro, Ariz....... | Oct. 3 | .... do .... | 2.03 | 2.35 | 0.40 | 0.70 |

PARIDÆ (the Titmice).

11. *Lophophanes inornatus* (Gamb.)—Gray-tufted Titmouse.

Common. Is found chiefly among the piñons and scrub-cedars, and also in Southern Arizona frequents the oaks. Usually goes in small flocks.

12. *Lophophanes wollweberi*, Bp.—Wollweber's Titmouse.

A small number was met with at Apache, Ariz., in a grove of scattered oaks, and in company with a large flock of the succeeding species.

13. *Parus montanus*, Gamb.—Mountain-Chickadee.

Frequently met with, chiefly among the heavy pines; also found among the oaks. Rarely seen in large flocks, but during the fall a few are certain to be seen accompanying each flock of warblers, nuthatches, &c. Habits and notes do not differ essentially from those of its congeners.

14. *Psaltriparus minimus* (Towns.), var. *plumbeus*, Bd.—Lead-colored Titmouse.

Not found at all in the heavy pines, but elsewhere abundant, and in fall and winter in very large flocks. One of the most active of the family; constantly on the move from tree to tree, searching for insects. Perfectly fearless and unsuspicious.

| No. | Sex. | Locality. | Date. | Collector. | Wing. | Tail. | Bill. | Tarsus. |
|---|---|---|---|---|---|---|---|---|
| 451 | ♀ | Fort Wingate, N. Mex. | July 16 | Henshaw. | 1.95 | 2.46 | 0.27 | 0.58 |
| 581 | ♀ jun. | Apache, Ariz ......... | Aug. 21 | .... do .... | 1.92 | 1.85 | 0.31 | 0.61 |
| 608 | ♂ | ......do ............... | Aug. 24 | .... do .... | 1.93 | 1.93 | 0.28 | 0.60 |

15. *Auriparus flaviceps* (Sund.)—Yellow-headed Titmouse; Verdin.

A single specimen was taken at old Camp Goodwin, Ariz., and occasionally an individual was met with among the mezquite-trees along the Gila River. Their habits, so far as observed, seemed to resemble those of the titmice, as do also its great variety of notes.

| No. | Sex. | Locality. | Date. | Collector. | Wing. | Tail. | Bill. | Tarsus. |
|---|---|---|---|---|---|---|---|---|
| 792 | ♂ | Camp Goodwin, Ariz.. | Sept. 17 | Henshaw. | 2.07 | 1.87 | 0.35 | 0.57 |
| 906 | ♂ | Gila River, Ariz ...... | Oct. 3 | ..., do .... | 1.90 | 1.93 | 0.33 | 0.60 |

SITTIDÆ (the Nuthatches).

16. *Sitta carolinensis* Gm., var. *aculeata*, Cass.—Slender-billed Nuthatch.

Common in the pineries.

17. *Sitta pusilla* Lath., var. *pygmæa*, Vig.—Pigmy Nuthatch.

Abundant. In summer exclusively pinicoline, but in the fall it is often seen in the groves of evergreen oaks. Gregarious, or nearly so, at all seasons. In the fall it gathers together in flocks of fifty or more, and, in company with the warblers and titmice, may be seen constantly on the move in search for food, when its loud *weet-weet* is continually emitted as it moves along the branches or takes flight.

| No. | Sex. | Locality. | Date. | Collector. | Wing. | Tail. | Bill. | Tarsus. |
|-----|------|-----------|-------|------------|-------|-------|-------|---------|
| 486 | ♀ jun. | Inscription Rock, N. Mex. | July 24 | Henshaw. | 2.55 | 1.55 | 0.55 | 0.57 |
| 988 | ♂ | Mountains, source of Gila River. | Nov. 5 | .... do .... | 2.62 | 1.75 | 0.57 | 0.63 |

CERTHIIDÆ (the Creepers).

18. *Certhia familiaris* L., var. *americana*, Bon.—Brown Creeper.

Perhaps not uncommon. I saw but few, and only among the pines of the mountains.

| No. | Sex. | Locality. | Date. | Collector. | Wing. | Tail. | Bill. | Tarsus. |
|-----|------|-----------|-------|------------|-------|-------|-------|---------|
| 670 | ♂ | Apache, Ariz.......... | Sept. 1 | Henshaw. | 2.63 | 2.80 | 0.64 | 0.60 |

TROGLODYTIDÆ (the Wrens).

19. *Campylorhynchus brunneicapillus* (Lafr.)—Cactus-Wren.

The region along the Gila River, in Arizona, seems eminently adapted to the habits of this bird, as the various species of cacti, the thickets of which it specially frequents, are nowhere more numerously represented than here. While along the river, the middle of September, though I searched carefully, I saw but a single individual. Its actions suggested those of the Winter-Wren. It had taken refuge in a dense thicket, from which its harsh, scolding notes could be heard as it passed from one part to another, keeping itself carefully hidden from view. Its curiosity finally inducing it to venture to the top to watch me, I secured it.

| Locality. | Date. | Collector. | Wing | Tail. | Bill. | Tarsus. |
|-----------|-------|------------|------|-------|-------|---------|
| Gila River, Ariz .............. | Sept. 15 | Henshaw ..... | 3.17 | 2.97 | 0.90 | 1.20 |

20. *Salpinctes obsoletus* (Say).—Rock-Wren.

This is an abundant species throughout Eastern Arizona, everywhere frequenting the masses of broken rocks, and showing an especial proclivity for those of volcanic nature. The young in nesting-plumage were taken at Wingate, N. Mex., July 14. On the 28th, at Zuni, a nest was found containing four young nearly fledged. The nest proper was merely a pile of grasses, slightly hollowed, and lined with horse-hairs and bits of sheep's wool. This was placed in a natural cavity of a clayey bank. It was without doubt a second brood.

21. *Catherpes mexicanus* (Swains.), var. *conspersus* Ridg.—White-throated Rock-Wren.

Not uncommon in vicinity of Apache, Ariz., and met with frequently from here to the southward; also found common in New Mexico near the sources of the Gila. This species is rarely seen but in the deep cañons and along the sides of rocky glens, and, like the former, is found most often among volcanic rocks. Its song is loud, clear, and melodious, and, once heard, is never to be mistaken for that of any other bird. It consists of a series of loud, detached whistles, which, beginning at a high note, descend smoothly and gradually through the entire scale.

| No. | Locality. | Date. | Collector. | Wing. | Tail. | Bill. | Tarsus. |
|---|---|---|---|---|---|---|---|
| 669 | Apache, Ariz.................. | Sept. 11 | Henshaw. | 2.40 | 2.45 | 0.76 | 0.73 |
| 986 | Mountains near Gila River, N. Mex. | Nov. 5 | .... do .... | 2.29 | 2.32 | 0.75 | 0.70 |
| 993 | ......do...................... | Nov. 5 | .... do .... | 2.25 | 2.07 | 0.75 | 0.72 |

22. *Thryothorus bewickii* (Aud.)

A rather common species in the neighborhood of Apache; also found at various points south. Undoubtedly occurs likewise in New Mexico, though I did not see it there. Prefers thickets and clumps of bushes on the open hill-sides, where it spends much of its time upon the ground searching for food.

| No. | Sex. | Locality. | Date. | Collector. | Wing. | Tail. | Bill. | Tarsus. |
|---|---|---|---|---|---|---|---|---|
| 590 | ♂ | Apache, Ariz ......... | Aug. 21 | Henshaw. | 2.25 | 2.15 | ...... | 0.80 |
| 600 | ♂ | ......do ............. | Aug. 23 | .... do .... | 2.25 | 2.35 | 0.58 | 0.73 |
| 750 | ♂ | Southern Arizona...... | Sept. 11 | .... do .... | 2.33 | 2.40 | 0.55 | 0.63 |

23. *Troglodytes aëdon* Vieill., var. *parkmanni* Aud.—Parkman's Wren.

Common everywhere wherever thickets, clumps of bushes, and fallen logs afford a good hunting-ground.

24. *Telmatodytes palustris* (Wils.), var. *paludicola* Baird —Long-billed Marsh-Wren.

A few seen at Apache in a small clump of reeds. Numerous in any marshy spot which is suited to its habits.

SYLVICOLIDÆ (the Warblers).

25. *Helminthophaga virginiæ*, Bd.—Virginia's Warbler.

Apparently a rather rare species in Arizona. I shot two specimens in a willow-thicket, by the side of a stream, in the White Mountains, August 11; not met with again.

| No. | Sex. | Locality. | Date. | Collector. | Wing. | Tail. | Bill. | Tarsus. |
|-----|------|-----------|-------|------------|-------|-------|-------|---------|
| 553 | ♀ | White Mts., Ariz...... | Aug. 11 | Henshaw. | 2.03 | 1.93 | 0.42 | 0.67 |
| 554 | ♀ | ......do ............... | Aug. 11 | ....do .... | 2.11 | 2.00 | 0.40 | 0.65 |

26. *Helminthophaga celata*, (Say.)—Orange-crowned Warbler.

Rather common, keeping generally in the low thickets and brush of the streams.

Bill and feet brown; soles yellow.

| No. | Sex. | Locality. | Date. | Collector. | Wing. | Tail. | Bill. | Tarsus. |
|-----|------|-----------|-------|------------|-------|-------|-------|---------|
| 662 | ♂ | Apache, Ariz.......... | Sept. 1 | Henshaw. | 2.43 | 2.06 | 0.38 | 0.66 |

27. *Dendroica æstiva*, (Gm.)—Yellow Warbler.

Common.

28. *Dendroica audubonii*, (Towns.)—Audubon's Warbler.

This species breeds quite commonly in the mountains of Southern Colorado, and I think it most probable that a few spend the summer in the higher portions of the White Mountains, Arizona, as I met with several here quite early in August. During the fall-migration, they were tolerably numerous, and were numerous October 17. At this season, they show no decided preference of locality, but are found in about equal numbers in the low scrub of the hill-sides and among the deciduous trees of the streams. They are adroit and successful fly-catchers.

| No. | Sex. | Locality. | Date. | Collector. | Wing. | Tail. | Bill. | Tarsus. |
|-----|------|-----------|-------|------------|-------|-------|-------|---------|
| 664 | ♂ ad. | Apache, Ariz ......... | Sept. 1 | Henshaw. | 3.22 | 2.58 | 0.40 | 0.74 |
| 972 | ♀ jun. | Gila River, N. Mex ... | Oct. 11 | ....do .... | 2.82 | 2.33 | 0.40 | 0.72 |
| 972a | ♀ | ......do·............... | Oct. 11 | ....do .... | 3.04 | 2.40 | 0.40 | 0.70 |

29. *Dendroica graciæ*, Cs.—Arizona Warbler.

An abundant Warbler in the White Mountains, Arizona, in August' where also it doubtless breeds. Numbers were observed accompanying flocks of chickadees, nuthatches, and other warblers. A single adult female, in worn breeding-plumage, was taken at Inscription Rock, July 24.

Iris black; bill and feet brown; soles light-yellow.

| No. | Sex. | Locality. | Date. | Collector. | Wing. | Tail. | Bill. | Tarsus. |
|-----|------|-----------|-------|------------|-------|-------|-------|---------|
| 485 | ♀ ad. | Insc'pt'n Rock, N. Mex. | July 24 | Henshaw. | 2.53 | 2.10 | 0.45 | 0.68 |
| 520 | ♂ ad. | Apache, Ariz.......... | Aug. 21 | ....do .... | 2.67 | 2.24 | 0.42 | 0.63 |
| 534 | ♂ ad. | White Mts., Ariz...... | Aug. 8 | ....do .... | 2.63 | 2.26 | 0.40 | 0.72 |
| 535 | ♀ jun. | ......do ............... | Aug. 9 | ....do .... | 2.50 | 2.27 | 0.40 | 0.60 |
| 567 | ♀ jun. | ......do ............... | Aug. 11 | ....do .... | 2.55 | 2.16 | 0.38 | 0.60 |
| 691 | ♂ jun. | Apache, Ariz..... .... | Sept. 3 | ....do .... | 2.63 | 2.30 | 0.40 | 0.64 |
| 747 | ♀ ad. | South Apache, Ariz ... | Sept. 3 | ....do .... | 2.60 | 2.15 | 0.39 | 0.60 |

30. *Dendroica nigrescens*, (Towns.)—Black-throated Gray Warbler.

Found to be quite common, August 12, in the pine-woods of the White Mountains. Probably breeds. In habits more active and restless than the warblers generally.

| No. | Sex. | Locality. | Date. | Collector. | Wing. | Tail. | Bill. | Tarsus. |
|---|---|---|---|---|---|---|---|---|
| 529 | ♂ ad. | Apache, Ariz.......... | Aug. 12 | Henshaw. | 2.50 | 2.17 | 0.38 | 0.67 |
| 566 | ♂ jun. | White Mts., Ariz...... | Aug. 12 | ....do .... | 2.45 | 2.25 | 0.40 | 0.64 |
| 578 | ♂ ad. | Apache, Ariz.......... | Aug. 21 | ....do .... | 2.50 | 2.17 | 0.40 | 0.69 |
| 601 | ♀ jun. | ......do ............ | Aug. 23 | ....do .... | 2.27 | 2.06 | 0.40 | 0.65 |
| 701 | ♂ ad. | ......do ............ | Sept. 5 | ....do .... | 2.35 | 2.17 | 0.38 | 0.69 |

31. *Geothlypis trichas*, (L.)—Maryland Yellowthroat.

At Apache I heard a male of this species in a thicket by the river-side. It is regarded by Dr. Coues as a rare summer-resident.

32. *Geothlypis macgillivrayi*, (Aud.)—Macgillivray's Warbler.

One of the most abundant and generally distributed of the warblers; found in every moist locality among the thickets.

| No. | Sex. | Locality. | Date. | Collector. | Wing. | Tail. | Bill. | Tarsus. |
|---|---|---|---|---|---|---|---|---|
| 123 | ♂ jun. | Bowie, Ariz.......... | Oct. 7 | Newberry. | 2.28 | 2.14 | 0.42 | 0.75 |
| 555 | ♀ | White Mts., Ariz...... | Aug. 11 | Henshaw . | 2.25 | 2.37 | 0.43 | 0.78 |
| 556 | ♂ jun. | ......do ............ | Aug. 11 | ....do .... | 2.20 | 2.36 | 0.42 | 0.93 |
| 661 | ♀ | Apache, Ariz.......... | Sept. 1 | ....do .... | 2.18 | 2.30 | 0.44 | 0.76 |
| 663 | ♀ | ......do ............ | Sept. 1 | ....do .... | 2.29 | 3.30 | 0.45 | 0.80 |

33. *Icteria virens* (L.), var. *longicauda*, Lawr.—Long-tailed Chat.

Met with but twice during the season; at Apache August 5, when I took a young male, just molting the nesting-plumage, and again a single bird seen at Wingate.

| No. | Sex. | Locality. | Date. | Collector. | Wing. | Tail. | Bill. | Tarsus. |
|---|---|---|---|---|---|---|---|---|
| 523 | ♂ jun. | Apache, Ariz.......... | Aug. 5 | Henshaw. | 2.93 | 3.38 | 0.55 | 0.97 |

34. *Myiodioctes pusillus*, (Wils.)—Green Black-capped Flycatcher.

One or two seen at Apache early in August, and by the first of September and during this month it was very abundant, much more so than I ever saw it at the East. It frequents chiefly the deciduous trees and bushes of the streams, and is found from the plains well up to the tops of the mountains.

| No. | Sex. | Locality. | Date. | Collector. | Wing. | Tail. | Bill. | Tarsus. |
|---|---|---|---|---|---|---|---|---|
| 671 | ♂ jun. | Apache, Ariz.......... | Sept. 1 | Henshaw . | 2.20 | 2.15 | 0.38 | 0.73 |
| 693 | ♀ jun. | ......do ............ | Sept. 3 | ....do .... | 2.16 | 2.16 | 0.40 | 0.65 |
| 757 | ♀ jun. | Thirty miles south of Apache, Ariz ...... | Sept. 3 | Maguet .. | 2.13 | 2.05 | 0.37 | 0.67 |
| 115 | ♂ ad. | Bowie, Ariz.......... | Oct. 6 | Newberry. | 2.17 | 2.15 | 0.35 | 0.66 |

35. *Setophaga picta*, Sw.—Painted Flycatcher.

At Apache, August 29, a specimen was presented to me by Lieutenant Manning, which was obtained in a bushy cañon immediately back of the fort. On leaving Apache and going south, from the 1st to the 15th of September, I saw perhaps a dozen individuals, and judge that the mountains of this section afford it a summer-home. The species has been noted at Tucson, Ariz., where it was seen migrating on two occasions in April and September by Captain Bendire. Their habits and motions are much like those of the Redstart (*S. ruticilla*). They frequented the bushes and smaller trees, especially the oaks. With half-shut wings and outspread tail, they pass rapidly along the limbs, now and then making a sudden dart for a passing fly, which secured they again alight and resume their search. They are constantly in motion, and rarely remain in the same tree many moments. It not infrequently may be seen clinging to the trunk of a tree, while it seizes a grub or minute insect which its sharp eyes have detected hidden in the bark.
Bill and feet black.

| No. | Sex. | Locality. | Date. | Collector. | Wing. | Tail. | Bill. | Tarsus. |
|---|---|---|---|---|---|---|---|---|
| 651 | ♂ jun. | Apache, Ariz.......... | Aug. 29 | Manning . | 2.70 | 2.55 | 0.36 | 0.60 |
| 732 | ♂ ad. | Thirty miles south of Apache, Ariz ...... | Sept. 10 | Henshaw. | 2.82 | 2.66 | 0.40 | 0.64 |
| 740 | ♂ ad. | ......do .............. | Sept. 10 | ....do .... | 2.73 | 2.64 | 0.38 | 0.61 |
| 741 | ...... | ......do .............. | Sept. 11 | ....do .... | 2.70 | 2.70 | 0.37 | 0.60 |

HIRUNDINIDÆ (the Swallows).

36. *Progne subis*, (L.)—Purple Martin.

Abundant, both in New Mexico and Arizona, generally at a considerable altitude. At Apache the parent birds were feeding the fledged young August 22.

| No. | Sex. | Locality. | Date. | Collector. | Wing. | Tail. | Bill. | Tarsus. |
|---|---|---|---|---|---|---|---|---|
| 524 | ♂ ad. | Apache, Ariz ......... | Aug. 5 | Henshaw. | 5.91 | 3.35 | 0.50 | 0.57 |
| 595 | ♂ jun. | ......do .............. | Aug. 22 | ....do .... | 4.55 | 2.27 | 0.41 | 0.60 |

37. *Petrochelidon lunifrons*, (Say.)—Cliff-Swallow.

A wide-spread species, both in Arizona and New Mexico, as their mud-nests, attached to the cliffs everywhere, attest.

38. *Hirundo horreorum*, Barton.—Barn-Swallow.

Noted by Dr. Newberry as rather common about Santa Fé, where several were shot; and also seen on the road between here and Fort Wingate.

39. *Tachycineta thalassina*, (Sw.)—Violet-green Swallow.

An abundant swallow near Wingate, N. Mex.; equally so in the

White Mountains, Arizona.   A rather exclusive inhabitant of the pine-woods, rarely descending to the plains.

| No. | Sex. | Locality. | Date. | Collector. | Wing. | Tail. | Bill. | Tarsus. |
|-----|------|-----------|-------|-----------|-------|-------|-------|---------|
| 453 | ♂ ad. | Xentria, N. Mex........ | July 19 | Henshaw. | 4.50 | 2.07 | 0.25 | 0.40 |

**40. Stelgidopteryx serripennis, (Aud.)—Rough-winged Swallow.**
Seen near Zuni, N. Mex. Found by Dr. Coues as an abundant summer-resident.

### VIREONIDÆ (the Vireos).

**41. Vireo gilvus (Vieill.), var. swainsoni, Bd.—Western Warbling Vireo.**
Numerous.

| No. | Sex. | Locality. | Date. | Collector. | Wing. | Tail. | Bill. | Tarsus. |
|-----|------|-----------|-------|-----------|-------|-------|-------|---------|
| 557 | ♂ ad. | White Mountains, Ariz. | Aug. 11 | Henshaw. | 2.67 | 2.15 | 0.43 | 0.65 |
| 558 | ♀ | .......do ............... | Aug. 11 | ....do .... | 2.50 | 1.95 | 0.45 | 0.66 |

**42. Vireo solitarius (Wils.), var. cassini, Bd.—Cassin's Vireo.**
A single specimen was shot a short distance south of Apache by Dr. Newberry, jr., in September. It appeared to be rather common on the Gila River, Arizona, keeping in the tops of the tall cottonwoods. I observed nothing peculiar in its habits.

| No. | Sex. | Locality. | Date. | Collector. | Wing. | Tail. | Bill. | Tarsus. |
|-----|------|-----------|-------|-----------|-------|-------|-------|---------|
| 92 | ♀ | South of Apache, Ariz. | Sept. 12 | Newberry. | 2.78 | 2.29 | 0.45 | 0.75 |
| 781 | ♀ | Gila River, Ariz ...... | Sept. 16 | Henshaw . | 2.82 | 2.25 | 0.45 | 0.60 |
| 795 | ...... | .......do ............... | Sept. 16 | ....do .... | 2.85 | 2.30 | 0.45 | 0.73 |

**43. Vireo solitarius (Wils.), var. plumbeus, Cs.—Western Solitary Vireo.**
Common at various places in New Mexico, and also in Eastern Arizona, keeping among the pines. A specimen, taken September 3, at Apache, Ariz., is intermediate in coloration between solitarius and plumbeus, showing very clearly the relationship of the two. A strong greenish tinge pervades the back, and is also very decided on the sides and flanks.

| No. | Sex. | Locality. | Date. | Collector. | Wing. | | Bill. | rsus . |
|-----|------|-----------|-------|-----------|-------|---|-------|--------|
| 461 | ♂ ad. | Neutria, N. Mex....... | July 19 | Henshaw. | 2.15 | 2.54 | 0.52 | 0.75 |
| 462 | ♂ ad. | .......do ............... | July 19 | ....do .... | 2.13 | | 0.55 | 0.74 |
| 694 | ♂ | Apache, Ariz........... | Sept. 3 | ....do .... | 2.17 | 2.48 2.45 | 0.51 | 0.74 |

**44. Vireo belli, Aud.—Bell's Vireo.**
This little vireo appeared to be rather common along the Gila River, inhabiting the dense thickets along the banks. At this season, the

middle of September, its quaint song was heard during most of the day, but more particularly in the hot hours of noonday. In addition to the song, which somewhat resembles the White-eyed Vireo's, it has a harsh scolding-note, which it often repeats as it searches among the dense undergrowth for its food. But a single specimen was obtained, as it was rather timid, and on hearing the slightest noise would instantly cease its notes and dive into the brush.

| No. | Sex. | Locality. | Date. | Collector. | Wing. | Tail. | Bill. | Tarsus. |
|------|------|-----------|-------|------------|-------|-------|-------|---------|
| 794 | ♂ ad. | Gila River, Ariz ...... | Sept. 16 | Henshaw. | 2.15 | 2.23 | 0.43 | 0.75 |

AMPELIDÆ (the Chatterers).

45. *Ampelis cedrorum*, (Vieill.)—Cedar-Bird.

The only one seen, during the entire season, was obtained in a small cañon, about thirty miles south of Apache, Ariz. The plumage was very much worn, and the bird, a female, had probably nested in the neighborhood.

Iris brown; bill and feet black.

| No. | Sex. | Locality. | Collector. | Wing. | Tail. | Bill. | Tarsus. |
|------|------|-----------|------------|-------|-------|-------|---------|
| 752 | ♀ ad. | South of Apache, Ariz .... | Henshaw. | 3.45 | 2.32 | 0.38 | 0.62 |

46. *Phænopepla nitens*, (Sw.)—Black Flycatcher.

Large numbers of this species were found, on several occasions, in the cañon back of Camp Apache. As they were noticed nowhere else in this vicinity, I judged that the abundance of mistletoe-berries here served as an attraction. These they were greedily feeding upon. In a cañon at the base of Mount Turnbull I also saw large numbers. Here the berries, which appear to be a favorite diet, were wanting, and they were engaged much of the time in catching flies, which they did by ascending perpendicularly from the bushes, snapping up an insect, and returning, much in the manner of the bluebirds. At this season they are very restless and shy.

| No. | Sex. | Locality. | Date. | Collector. | Wing. | Tail. | Bill. | Tarsus. |
|------|------|-----------|-------|------------|-------|-------|-------|---------|
| 515 | ♂ jun. | Apache, Ariz .......... | Aug. 4 | Henshaw. | 3.35 | 3.37 | 0.45 | 0.61 |
| 516 | ♂ jun. | .......do ............ | Aug. 4 | ....do .... | 3.42 | 3.42 | 0.45 | 0.63 |
| 797 | ♂ ad. | Mount Turnbull, Ariz. | Aug. 17 | ....do .... | 3.48 | 3.58 | 0.45 | 0.63 |
| 800 | ♂ jun. | .......do ............ | Sept. 19 | ....do .... | 3.75 | 4.12 | 0.40 | 0.70 |

47. *Myiadestes townsendii*, (Aud.)—Townsend's Solitaire.

Quite common, in the fall, in Eastern Arizona and Western New Mexico. Having reared their young, these birds appear to forsake the pine-woods, which constitute their summer-abode, and appear lower down on the hill-sides, covered with piñon and cedars. Their food at this season appears to consist almost exclusively of berries, particularly from the piñon and cedars, and the crops of many examined contained little else save a few insects. The habit of catching insects on the wing, after the manner of the flycatchers, which is attributed to this bird, appears to be not a common one, as, of hundreds I have seen at different seasons, none were ever thus engaged, nor have I ever seen them searching among the leaves for insects, like the thrushes. In their usual manner of procuring food, as in their habits and motions generally, they have always seemed to me nearly allied to the bluebirds. Though in summer a bird of retiring and unsocial habits, and never more than a single pair being found in a locality, in the fall they are to a considerable extent gregarious, associating usually in small companies of from five to ten. At the Old Crater, forty miles south of Zuni, N. Mex., they had congregated in very large numbers about a spring of fresh water, the only supply for many miles around, and hundreds were to be seen sitting on the bare volcanic rocks, apparently too timid to venture down and slake their thirst while we were camped near by. Their song is occasionally heard even in November and December, and is very sweet, but not so full and varied as during the vernal season.

| No. | Sex. | Locality. | Date. | Collector. | Wing. | Tail. | Bill. | Tarsus. |
|---|---|---|---|---|---|---|---|---|
| 453 | ♂ ad. | Wingate, N. Mex...... | July 18 | Henshaw. | 4.50 | 4.21 | 0.48 | 0.75 |
| 704 | ♀ ad. | Apache, Ariz ......... | Sept. 6 | ....do .... | 4.40 | 3.97 | 0.47 | 0.78 |
| 959 | ♂ ad. | Silver City, N. Mex ... | Oct. 24 | ....do .... | 4.50 | 4.10 | 0.45 | 0.81 |
| 960 | ♂ | ......do .......... .... | Oct. 24 | ....do .... | 4.57 | 4.32 | 0.49 | 0.80 |
| 961 | ♂ | ......do .......... .... | Oct. 24 | ....do .... | 4.56 | 4.41 | 0.45 | 0.75 |

LANIIDÆ (the Shrikes).

48. *Collurio ludovicianus* (L.), var. *excubitoroides*, Sw.—White-rumped Shrike.

A single individual seen at Wingate, N. Mex., in July, and a specimen obtained by Dr. Newberry at Fort Bowie, Southeastern Arizona, were the only two occasions when this species was met with. Dr. Coues mentions this shrike as rare also at Fort Whipple.

| No. | Sex. | Locality. | Date. | Collector. | Wing. | Tail. | Bill. | Tarsus. |
|---|---|---|---|---|---|---|---|---|
| 116 | ♂ ad. | Bowie, Ariz .......... | Oct. 6 | Newberry. | 2.55 | 4.15 | 0.60 | 1.02 |

TANAGRIDÆ (the Tanagers).

49. *Pyranga ludoviciana*, (Wils.)—Louisiana Tanager.

Very common at Apache, and met with frequently at various points to the southward. Seen at the Gila River October 16, but at this time nearly all had gone farther south. Frequents at this season the deciduous trees.

Iris brown; bill horn-color; feet and legs bluish.

| No. | Sex. | Locality. | Date. | Collector. | Wing. | Tail. | Bill. | Tarsus. |
|-----|------|-----------|-------|------------|-------|-------|-------|---------|
| 460 | ♀ ad. | Neutria, N. Mex ...... | July 19 | Henshaw | 3.93 | 2.90 | 0.58 | 0.73 |
| 585 | ♂ ad. | Apache, Ariz.......... | Aug. 21 | .... do .... | 3.68 | 3.07 | 0.65 | 0.75 |
| 627 | ♀ jun. | ....do ................ | Aug. 27 | .... do .... | 3.70 | 3.18 | 0.63 | 0.83 |
| 709 | ♂ ad. | ....do ................ | Sept. 7 | .... do .... | 3.75 | 3.15 | 0.64 | 0.74 |
| 742 | ♀ jun. | South of Apache, Ariz. | Sept. 11 | .... do .... | 3.68 | 2.98 | 0.63 | 0.75 |
| 745 | ♀ jun. | ....do ................ | Sept. 11 | .... do .... | 3.55 | 2.88 | 0.59 | 0.74 |
| 746 | ♀ jun. | ....do ................ | Sept. 11 | .... do .... | 3.78 | 2.90 | 0.64 | 0.80 |
| 799 | ♀ ad. | Goodwin, Ariz ........ | Sept. 17 | .... do .... | 3.62 | 2.86 | 0.58 | 0.79 |
| 915 | ♀ jun. | Gila River, Ariz ...... | Oct. 16 | .... do .... | 3.70 | 2.94 | 0.65 | 0.80 |

50. *Pyranga hepatica*, Sw.—Liver-colored Tanager.

A single female of this little-known species was shot at Apache August 4. In a grove of oaks on the skirts of a pine-forest, about twenty miles south of Apache, I saw, in the course of an afternoon, perhaps half a dozen males. They appeared to be feeding upon insects, which they gleaned from among the foliage and smaller branches of the oaks. They were excessively shy, so much so that I found it difficult to get within gun-shot of them. They probably spend the summer in the mountains, at least as far north as Apache. The species was introduced into our fauna by Dr. Woodhouse, who took a single female in the San Francisco Mountains, New Mexico. No other specimens have since been obtained till the present time.

| No. | Sex. | Locality. | Date. | Collector. | Wing. | Tail. | Bill. | Tarsus. |
|-----|------|-----------|-------|------------|-------|-------|-------|---------|
| 511 | ♀ ad. | Apache, Ariz.......... | Aug. 4 | Henshaw. | 4.05 | 3.05 | 0.68 | 0.83 |
| 717 | ♂ ad. | Twenty miles south of Apache, Ariz. | Sept. 8 | .... do .... | 4.10 | 3.53 | 0.67 | 0.87 |
| 718 | ♂ ad. | ....do ................ | Sept. 8 | .... do .... | 4.11 | 3.32 | 0.68 | 0.85 |

51. *Pyranga æstiva* (L.), var. *cooperi*, Ridg.—Cooper's Tanager.

A beautiful adult male of this variety of the Summer-Tannager (*P. æstiva*) was taken on the Gila River, Arizona, September 16, and another heard in same locality; also noted on the San Francisco River October 10. In each instance they were found in the tall cottonwoods, actively engaged searching for insects.

| No. | Sex. | Locality. | Date. | Collector. | Wing. | Tail. | Bill. | Tarsus. |
|-----|------|-----------|-------|------------|-------|-------|-------|---------|
| 792 | ♂ ad. | Gila River, Ariz ...... | Sept. 16 | Henshaw. | 4.00 | 3.50 | 0.83 | 0.80 |

FRINGILLIDÆ (the Finches).

52. *Hesperiphona vespertina*, (Coop.)—Evening-Grossbeak.

A small flock of immature birds were seen a little south of Apache feeding upon savis-berries. Not seen elsewhere.

| No. | Sex. | Locality. | Date. | Collector. | Wing. | Tail. | Bill. | Tarsus. |
|-----|------|-----------|-------|------------|-------|-------|-------|---------|
| 739 | ♀ jun. | South of Apache, Ariz. | Sept. 11 | Henshaw. | 4.14 | 2.64 | 0.70 | 0.75 |

53. *Carpodacus cassini*, Bd.—Cassin's Purple Finch.
A large flock seen at the salt lake south of Zuni November 20.

| No. | Sex. | Locality. | Date. | Collector. | Wing. | Tail. | Bill. | Tarsus. |
|---|---|---|---|---|---|---|---|---|
| 999 | ♂ ad. | Salt Lake, south Zuni | Nov. 20 | Henshaw. | 3.56 | 2.55 | 0.51 | 0.74 |
| 1000 | ♂ ad. | ....do .............. | Nov. 20 | .... do .... | 3.55 | 2.54 | 0.56 | 0.74 |
| 1001 | ♂ ad. | ....do .............. | Nov. 20 | .... do .... | 3.83 | 2.92 | 0.55 | 0.75 |
| 1002 | ♂ jun. | ....do .............. | Nov. 20 | .... do .... | 3.68 | 2.67 | 0.52 | 0.70 |
| 1003 | ♀ | ....do .............. | Nov. 20 | .... do .... | 3.50 | 2.72 | 0.54 | 0.70 |
| 1004 | ♂ jun. | ....do .............. | Nov. 20 | .... do .... | 3.53 | 2.65 | 0.52 | 0.70 |
| 1005 | ♀ | ....do .............. | Nov. 20 | .... do .... | 3.62 | 2.70 | 0.52 | 0.70 |
| 1007 | ♂ | ....do .............. | Nov. 20 | .... do .... | 3.52 | 2.50 | 0.52 | 0.73 |
| 1008 | ♀ ad. | ....do .............. | Nov. 20 | .... do .... | 3.43 | 2.50 | 0.50 | 0.73 |

54. *Carpodacus frontalis*, (Say.)—House-Finch; Burion.
Noted at Santa Fé by Dr. Newberry, jr.; and it was not uncommon
at Apache, Ariz.

| No. | Sex. | Locality. | Date. | Collector. | Wing. | Tail. | Bill. | Tarsus. |
|---|---|---|---|---|---|---|---|---|
| 613 | ♂ jun. | Apache, Ariz ......... | Aug. 25 | Henshaw. | 3.12 | 2.63 | 0.40 | 0.65 |

55. *Chrysomitris psaltria*, (Say.)—Arkansas Finch.
Rather numerous in Western New Mexico and Eastern Arizona, in
summer, at least as far south as the thirty-third parallel.

| No. | Sex. | Locality. | Date. | Collector. | Wing. | Tail. | Bill. | Tarsus. |
|---|---|---|---|---|---|---|---|---|
| 478 | ♀ ad. | Inscription Rock, N. Mex. | Aug. 23 | Henshaw. | 2.50 | 1.80 | 0.37 | 0.50 |
| 477 | ♂ ad. | ....do.............. | Aug. 23 | .... do .... | 2.45 | 1.83 | 0.38 | 0.53 |
| 710 | ♂ ad. | Apache, Ariz.......... | Sept. 7 | .... do .... | 2.50 | 1.85 | 0.40 | 0.50 |
| 773 | ♀ ad. | Gila River, Ariz....... | Sept. 14 | .... do .... | 2.45 | 1.75 | 0.39 | 0.50 |

56. *Chrysomitris psaltria* (Say), var. *arizonæ*, Cs.—Arizona Goldfinch.
But a single specimen taken on the Gila River September 14. During
the summer this is probably the prevailing form in Southern Arizona
and New Mexico.

| No. | Sex. | Locality. | Date. | Collector. | Wing. | Tail. | Bill. | Tarsus. |
|---|---|---|---|---|---|---|---|---|
| 768 | ♂ ad. | Gila River, Ariz....... | Sept. 14 | Henshaw. | 2.50 | 1.80 | 0.85 | 0.48 |

57. *Plectrophanes ornatus*, (Towns.)—Chestnut-colored Bunting.
From the 29th of September, when this species was first noted at
Camp Grant, Ariz., till November 10, near Tulerosa, N. Mex., it was
frequently seen in large flocks on the dry, arid plains and plateaus of

Southeastern Arizona and Southwestern New Mexico. They move about in companies of hundreds, and when on the ground run nimbly among the grasses searching for seeds and insects. When approached, the whole flock squats silently among the herbage, and remains so quiet, and their colors blend so nicely with the surrounding tints, that it is almost impossible to detect them, though but a few feet distant. On taking wing, each bird emits a number of short, quavering chirps, which they repeat constantly as long as on the wing. Their flight is erratic and wild, and, once startled, they are apt to keep on the wing a long time, flying hurriedly about. I have occasionally seen a flock start from the ground, and, after circling excitedly about, suddenly start off in a straight line till nearly out of sight, and then, as if urged by some new impulse, suddenly wheel about and take a direct course back, alighting within a few feet from the starting-point.

| No. | Sex. | Locality. | Date. | Collector. | Wing. | Tail. | Bill. | Tarsus. |
|---|---|---|---|---|---|---|---|---|
| 128 | ♀ | Bowie, Ariz.............. | Oct. 10 | Newberry. | 3.04 | 2.23 | 0.43 | 0.75 |
| 129 | ♀ | ....do ................. | Oct. 10 | .... do .... | 3.28 | 2.49 | 0.41 | 0.86 |
| 892 | ♀ | Camp Grant, Ariz..... | Sept. 29 | .... do .... | 3.15 | 2.34 | 0.42 | 0.75 |
| 895 | ♀ | ....do ................. | Sept. 29 | .... do .... | 3.17 | 2.35 | 0.40 | 0.75 |
| 896 | ♀ | ....do ................. | Sept. 29 | .... do .... | 3.20 | 2.40 | 0.45 | 0.75 |
| 897 | ♀ | ....do ................. | Sept. 29 | .... do .... | 3.00 | 2.26 | 0.40 | 0.78 |
| 901 | ♂ | San Pedro, Ariz....... | Oct. 3 | .... do .... | 3.32 | 2.50 | 0.43 | 0.75 |
| 902 | ♂ | ....do ................. | Oct. 3 | .... do .... | 3.27 | 2.37 | 0.41 | 0.80 |
| 904 | ♂ | ....do ................. | Oct. 3 | .... do .... | 3.20 | 2.30 | 0.40 | 0.75 |
| 923 | ♂ | Gila River, Ariz ...... | Oct. 17 | .... do .... | 3.25 | 2.40 | 0.47 | 0.75 |
| 924 | ♀ | ....do ................. | Oct. 17 | .... do .... | 3.95 | 2.23 | 0.42 | 0.75 |
| 939 | ♂ | ....do ................. | Oct. 17 | .... do .... | 3.12 | 2.40 | 0.40 | 0.74 |
| 989 | ♂ | ....do ................. | Oct. 17 | .... do .... | 3.25 | 2.45 | ...... | ...... |

Iris brown; bill plumbeous-brown above, lighter beneath; feet dusky-brown.

58. *Plectrophanes maccownii*, Lawr.—Chestnut-shouldered Longspur.

Found throughout much the same region as the preceding, and with very similar habits.

| No. | Sex. | Locality. | Date. | Collector. | Wing. | Tail. | Bill. | Tarsus. |
|---|---|---|---|---|---|---|---|---|
| 130 | ♀ jun. | Bowie, Ariz........... | Oct. 10 | Newberry. | 3.17 | 2.10 | 0.45 | 0.70 |
| 919 | ♀ | Gila River, Ariz....... | Oct. 16 | Henshaw . | 3.37 | 2.18 | 0.46 | 0.72 |
| 952 | ♂ | Bayard, N. Mex....... | Oct. 22 | .... do .... | 3.45 | 2.40 | 0.46 | 0.75 |
| 954 | ♂ | ....do ................. | Oct. 22 | .... do .... | 3.43 | 2.37 | 0.50 | 0.75 |
| 955 | ♂ | ....do ................. | Oct. 22 | .... do .... | 3.37 | 2.25 | 0.46 | 0.69 |
| 956 | ♀ | ....do ................. | Oct. 22 | .... do .... | 3.30 | 2.28 | 0.45 | 0.75 |
| 958 | ♀ | ....do ................. | Oct. 22 | .... do .... | 3.65 | 2.60 | 0.45 | 0.76 |
| 950 | ♀ | ....do ................. | Oct. 22 | .... do .... | 3.40 | 2.30 | 0.47 | 0.75 |

59. *Centronyx bairdii*, (Aud.)—Baird's Sparrow.

The interesting fact of the discovery of Baird's Bunting in large numbers in Northern Dakota has been announced by Dr. Coues. Additional light is thrown upon the range of this almost unknown species by its discovery in Southeastern Arizona and Southwestern New Mexico. I found it in immense numbers, from September 20 till late in October, throughout the rolling plains along the bases of the mountains, and even quite high up among the foot-hills. It was usually associated with the Savanna and Yellow-winged Sparrows, and seems to embrace in its habits certain characteristics of either species. Its flight is particularly like that of the former bird, but even more wild and irregular. It pursues its zigzag course for a couple of hundred yards, and then, suddenly turning sharply to one side, alights behind some friendly bush or tuft of grass. Like the Yellow-winged Sparrow, it is difficult to flush, but seeks rather to evade search by running nimbly through the grass, changing its course frequently, and hiding wherever possible, flying only when hard pressed. A large number of specimens were secured, all molting, and many in extremely ragged plumage. From their condition, it is presumed that they were not migrants, but breed in the immediate locality. The following measurements, taken from fresh specimens, were selected from a series of over thirty :

| No. | Sex. | Locality. | Date. | Collector. | Length. | Stretch. | W. | T. | B. | Tars. |
|---|---|---|---|---|---|---|---|---|---|---|
| 812 | ♂ | Camp Grant, Ariz. | Sept. 22 | Henshaw. | 5.49 | 9.37 | 3.00 | 2.31 | 0.43 | 0.80 |
| 813 | ♂ | ....do........ | Sept. 22 | .... do .... | 5.74 | 9.25 | 2.93 | 2.25 | 0.45 | 0.79 |
| 814 | ♀ | ....do........ | Sept. 22 | .... do .... | 5.43 | 8.80 | 2.74 | 2.19 | 0.45 | 0.80 |
| 815 | ♂ | ....do........ | Sept. 22 | .... do .... | 5.43 | 9.25 | 3.00 | 2.37 | 0.41 | 0.79 |
| 816 | ♀ | ....do........ | Sept. 22 | .... do .... | 5.49 | 9.13 | 2.74 | 2.25 | 0.45 | 0.77 |
| 817 | ♂ | ....do........ | Sept. 22 | .... do .... | 5.62 | 9.19 | 3.00 | 2.25 | 0.43 | 0.80 |
| 818 | ♀ | ....do........ | Sept. 22 | Maqnet... | 5.37 | 8.80 | 2.62 | 2.12 | 0.45 | 0.77 |
| 819 | ♀ | ....do........ | Sept. 23 | Henshaw. | 5.49 | 8.74 | 2.62 | 2.12 | .... | .... |
| 821 | ♂ | ....do........ | Sept. 23 | .... do .... | 5.68 | 9.06 | 3.00 | 2.17 | 0.45 | 0.85 |
| 822 | ♂ | ....do........ | Sept. 23 | .... do .... | 5.46 | 9.06 | 2.80 | 2.12 | 0.45 | 0.82 |
| 823 | ♂ | ....do........ | Sept. 23 | .... do .... | 5.61 | 9.37 | 3.00 | 2.37 | 0.44 | 0.82 |
| 835 | ♂ | ....do........ | Sept. 23 | .... do .... | 5.74 | 9.37 | 3.06 | 2.30 | 0.43 | 0.83 |
| 836 | ♂ | ....do........ | Sept. 23 | .... do .... | 5.66 | 9.25 | 3.00 | 2.25 | 0.47 | 0.81 |
| 837 | ♂ | ....do........ | Sept. 23 | .... do .... | 5.48 | 9.43 | 2.80 | 2.18 | 0.47 | 0.82 |
| 838 | ♂ | ....do........ | Sept. 23 | .... do .... | 5.36 | 9.31 | 2.80 | 2.25 | 0.43 | 0.77 |
| 839 | ♂ | ,...do........ | Sept. 23 | .... do .... | 5.66 | 9.06 | 2.74 | 2.18 | 0.45 | 0.81 |
| 804 | ♂ | Mt. Graham, Ariz. | Sept. 21 | .... do .... | ........ | ........ | 2.80 | 2.27 | 0.47 | 0.83 |
| 807 | ♂ | ....do........ | Sept. 21 | .... do .... | ........ | ........ | 2.75 | 2.30 | 0.44 | 0.83 |
| 920 | ♀ | Gila River, N. Mex. | Oct. 16 | .... do .... | ........ | ........ | 2.65 | 1.95 | 0.46 | 0.81 |

60. *Passerculus savanna* (Wil.), var. *alaudinus*, Bp.—Western Savanna Sparrow.

During the fall often found on the high dry plateaus. Always numerous in the vicinity of sloughs and streams.

| No. | Sex. | Locality. | Date. | Collector. | Wing. | Tail. | Bill. | Tarsus. |
|---|---|---|---|---|---|---|---|---|
| 608 | ♀jun. | Mt. Graham, Ariz..... | Aug. 21 | Henshaw. | 2.60 | 2.20 | 0.40 | 0.73 |
| 620 | ♂ | Apache, Ariz........... | Aug. 26 | ....do .... | 2.75 | 2.23 | 0.42 | 0.80 |
| 634 | ♀jun. | ......do ............. | Aug. 27 | ....do .... | 2.50 | 2.10 | 0.43 | 0.75 |
| 665 | ♂jun. | ......do ............. | Aug. 23 | ....do .... | 2.65 | 2.26 | 0.43 | 0.82 |
| 847 | ...... | Camp Grant, Ariz..... | Aug. 23 | ....do .... | 2.75 | 2.26 | 0.45 | 0.75 |
| 894 | ♀ | ......do ............. | Sept. 29 | ....do .... | 2.60 | 2.10 | 0.42 | 0.77 |

61. *Poocætes gramineus* (Gm.), var. *confinis*, Bd.—Western Grass-Finch.

This and the preceding species are perhaps the most common and generally distributed in the West of the sparrow-tribe. They both frequent much the same localities, but the Grass-Finch is more constantly found on the dry plains, and entirely away from the vicinity of water.

| No. | Sex. | Locality. | Date. | Collector. | Wing. | Tail. | Bill. | Tarsus. |
|---|---|---|---|---|---|---|---|---|
| 11 | ♂ | Camp Grant, Ariz..... | Sept. 22 | Magnet... | 2.93 | 2.67 | 0.45 | 0.78 |
| 12 | ♂ | ......do .............. | Sept. 22 | ....do .... | 3.13 | 2.71 | 0.45 | 0.85 |
| 784 | ♂ | Gila River, Ariz....... | Sept. 15 | Henshaw. | 3.30 | 2.65 | 0.45 | 0.83 |

62. *Coturniculus passerinus* (Wils.), var. *perpallidus*, Ridg.—Western Yellow-winged Sparrow.

Found abundantly over the same area as the Baird's Bunting. The specimens obtained are all typical of this race, and differ very decidedly from the eastern form (*passerinus*) in the general predominance of the light tints through the entire plumage.

| No. | Sex. | Locality. | Collector. | Wing. | Tail. | Bill. | Tarsus. |
|---|---|---|---|---|---|---|---|
| 6 | ♂ | Southern Arizona.............. | Magnet .. | 2.43 | 2.13 | 0.48 | 0.76 |
| 13 | ♂ | Camp Grant, Ariz............. | .... do .... | 2.50 | 2.09 | 0.47 | 0.77 |
| 10 | ♂ | Mount Graham, Ariz........... | .... do .... | 2.40 | 2.05 | 0.50 | 0.75 |
| 610 | ♂ | ......do ..................... | Henshaw . | 2.50 | 2.06 | 0.50 | 0.76 |
| 612 | ♂ | ......do ..................... | .... do .... | 2.50 | 2.07 | 0.43 | 0.73 |
| 613 | ♀ | ......do ..................... | .... do .... | 2.40 | 2.10 | 0.47 | 0.71 |
| 623 | ♂ | ......do ..................... | .... do .... | 2.60 | 2.15 | 0.47 | 0.70 |
| 629 | ♀ | ......do ..................... | .... do .... | 2.45 | 2.05 | 0.46 | 0.70 |
| 630 | ♀ | ......do ..................... | .... do .... | 2.62 | 2.23 | 0.45 | 0.73 |
| 627 | ♂ | ......do ..................... | .... do .... | 2.50 | 2.08 | 0.49 | 0.76 |
| 782 | ♂ | Gila River, Ariz ............. | .... do .... | 2.55 | 2.10 | 0.55 | 0.73 |

63. *Zonotrichia leucophrys*, (Forst.)—White-crowned Sparrow.

Exceedingly abundant in the fall in the valleys of the San Pedro and Gila Rivers, Arizona. Frequents the bushes, more particularly the willows along the small streams.

Iris brown; bill above dusky-brown, below lighter; feet and legs light-brown.

| No. | Sex. | Locality. | Date. | Collector. | Wing. | Tail. | Bill. | Tarsus. |
|---|---|---|---|---|---|---|---|---|
| 120 | ♀ ad. | Bowie, Ariz........... | Oct. 6 | Newberry. | 2.95 | 3.00 | 0.46 | 0.88 |
| 125 | ♀ ad. | ......do .............. | Oct. 10 | ....do .... | 3.12 | 3.11 | 0.44 | 0.85 |

64. *Zonotrichia leucophrys* (Forst.), var. *intermedia*, Ridg.—Gambel's Finch.

Arrived from the north rather later than the preceding. Found equally abundant in same localities, and associating together.

| No. | Sex. | Locality. | Date. | Collector. | Wing. | Tail. | Bill. | Tarsus. |
|---|---|---|---|---|---|---|---|---|
| 93 | ♀ jun. | Pueblo Viejo, N. Mex. | Sept. 18 | Newberry. | 2. 94 | 2. 85 | 0. 42 | 0. 83 |
| 122 | ♀ jun. | Bowie, Ariz | Oct. 7 | ....do .... | 2. 88 | 3. 00 | 0. 44 | 0. 86 |
| 790 | ♂ | Gila River, Ariz | Sept. 16 | Henshaw . | 2. 90 | 2. 92 | 0. 45 | 0. 82 |
| 950 | ♂ jun. | Bayard, N. Mex | Oct. 19 | ....do .... | 2. 90 | 2. 97 | 0. 42 | 0. 84 |
| 950a | | .......do ......... | Oct. 19 | ....do .... | 2. 80 | 2. 98 | 0. 42 | 0. 83 |

### Synopsis of the genus Junco.

Common characters: Prevailing color plumbeous; the abdomen, crissum, and lateral tail-feathers white:

A. Ash of the jugulum with its posterior surface concave, and abruptly defined against the white of abdomen; sides tinged with ash; upper parts pure ash :
  1. *hyemalis*
  2. var. *aikeni.*

B. Jugulum abruptly defined against the white of abdomen, but convex; sides pinkish; dorsal region dark rufous-brown :
  1. *oregonus.*
  2. var. *annectens.*

C. Back bright-rufous :
  1. *caniceps.*
  2. var. *dorsalis.*
  3. var. *cinereus.*
  4. var. *alticola.*

By the above arrangement the group is divided into three distinct species, each having a single variety in the United States, while to *caniceps* as varieties are referred, though somewhat doubtfully, the extreme southern forms *cinereus* and *alticola*. *Hyemalis* of the Eastern Province is represented in the high northern Rocky Mountains (?) by the variety *aikeni*, distinguished by its larger size, the white bands of the wings, the greater amount of white on the tail-feathers, and the generally paler coloration, features all readily traceable to the effects of its cold alpine habitat. *Annectens*, also inhabiting the northern Rocky Mountains, is referable to *oregonus* of the Pacific coast, which it resembles in the fulvous sides, and in the dark rufous-brown of the dorsal region ; features peculiar to these two forms. From it, it is separable as a variety, by much the same differences, though less in degree, that exist between *hyemalis* and *aikeni*, differences assignable, too, to the same causes. It is larger, with paler colors throughout, having the plumbeous-black of *oregonus* replaced by a light ash, and also, as Mr. Aiken informs me, not infrequently shows a decided tendency to the white banding of the wings. This is well shown in a specimen taken at Fountain, Col., in December, which has two well-defined bands, though not quite so conspicuous as in typical examples of *aikeni*. *Junco caniceps* of the central Rocky Mountains of the United States is at once distinguished from any of the above by the bright, reddish, chestnut-brown of the interscapular region. In the southern Rocky Mountains in New Mexico and Arizona, is found var. *dorsalis*, which seems to combine certain features peculiar to both *caniceps* and *cinereus*, and also in certain other points to differ from either. In the restriction of red to the interscapular region it is like *caniceps ;* but in quite a number of specimens collected in New Mexico during the past season the tertiaries are strongly tinged with rufous, showing in this respect an approach to
8 o s

*cinereus*, where the chestnut of the back extends over the wing-coverts and inner secondaries. The bill above is brownish-black, below whitish, thus differing from *caniceps*, which has a flesh-colored bill, and apparently approaching *cinereus*, where it is black above, below yellow. Like *cinereus*, also, the pale ash of the throat fades gradually into the white of abdomen, instead of being, as in *caniceps*, abruptly defined. Of quite a large series of specimens collected by myself the past season, and others in the Smithsonian collection, I have seen none which are not readily assignable to one variety or the other by the distinctive features pointed out. The theory of hybridization, which might be admissible were only one or two specimens known possessing intermediate characters, seems wholly inadequate as an explanation in the case of either *annectens* or *dorsalis*, where the forms extend over very extensive regions, and preserve their distinctive characteristics intact. Whether *cinereus* of the table-lands of Mexico, with a local variety, *alticola* of the mountains of Guatemala, may not justly be entitled to specific rank, is a matter of considerable doubt. While the typical forms of *caniceps* and *cinereus* are widely different, *dorsalis*, intermediate in its habitat, seems also intermediate in its characters, and it therefore may be best to treat the two (*caniceps et cinereus*) as only separable as varieties rather than as distinct species. A large *suite*, however, of these birds collected in Mexico, which at present is wanting, might shed more light on the subject.

65. *Junco hyemalis* (L.), var. *aikeni*, Ridg.—White-winged Snowbird.

This race of the common snow-bird (*hyemalis*) is found late in the fall and winter, distributed over quite a large area in the middle Rocky Mountains of the United States. I found it and the two succeeding forms, mingled indiscriminately in large flocks, in El Paso County, Colorado, the middle of December. Mr. Aiken has had abundant opportunity to note the time and manner of its migrations, and from these it seems pretty certain that it finds its summer-home very far to the northward. According to Mr. Aiken, the first stragglers from the north do not make their appearance till about the 5th of October, and then in gradually increasing numbers till the 1st of December, when they come in large flocks, the last to arrive being the old and fully-plumaged males. While many of the females and young birds proceed farther to the south, the greater number of the adult males winter at some point farther to the north than El Paso County, as of the whole number seen during the winter only about two-fifths are males. Early in February the old birds begin to start northward, the general migration being delayed about a month. The habits of this race do not differ from those of its congeners.

66. *Junco oregonus*, (Towns.)—Oregon Snowbird.

Found abundantly in the neighborhood of Bayard, N. Mex., and generally distributed from this point northward, keeping in the low foot-hills and along the streams on the plains. Mr. Aiken informs me that comparatively but few of this species remain during the winter in his section, the greater proportion passing on still further to the south.

| No. | Sex. | Locality. | Date. | Collector. | Wing. | Tail. | Bill. | Tarsus. |
|-----|------|-----------|-------|-----------|-------|-------|-------|---------|
| 947 | ♀ | Bayard, N. Mex....... | Oct. 19 | Henshaw. | 3.07 | 2.78 | 0.43 | 0.78 |

**67. *Junco oregonus* (Towns.), var. *annectens*, Bd.—Pink-sided Snowbird.**

Numerous in El Paso County, Colorado, in December. Considerable numbers winter here, although, from the fact that a large majority of these are males, Mr. Aiken is led to believe that the greater number spend the winter farther south. I met with it near Silver City, Southwestern New Mexico, late in October, but it was not common.

| No. | Sex. | Locality. | Date. | Collector. | Wing. | Tail. | Bill. | Tarsus. |
|---|---|---|---|---|---|---|---|---|
| 963 | ♂ jun. | Silver City, N. Mex ... | Oct. 24 | Henshaw. | 3.43 | 3.18 | 0.48 | 0.79 |

**68. *Junco caniceps* (Woodh.), var. *dorsalis*, Henry.**

From near Wingate, N. Mex., where this variety was first seen the middle of July to the southward, it appears to entirely replace the true *caniceps*. It was very abundant, keeping generally well up among the mountains, even in November not appearing to straggle far down. The young, in nesting-plumage, taken July 19, showed in the coloration of the back and bill the peculiar features of the adult birds.

Bill brownish-black above, below whitish; legs and feet brown.

| No. | Sex. | Locality. | Date. | Collector. | Wing. | Tail. | Bill. | Tarsus. |
|---|---|---|---|---|---|---|---|---|
| 459 | ♂ jun. | Neutria, N. Mex....... | July 19 | Henshaw . | 3.00 | 3.18 | 0.43 | 0.78 |
| 544 | ♂ ad. | White Mountains, Ariz. | Aug. 10 | ....do .... | 3.20 | 3.18 | 0.43 | 0.86 |
| 673 | ♂ | ......do .............. | Sept. 1 | ....do .... | 3.25 | 3.23 | 0.43 | 0.79 |
| 683 | ♂ | ......do .............. | Sept. 2 | ....do .... | 3.12 | 3.06 | 0.43 | 0.86 |
| 684 | ♀ | ..... do .............. | Sept. 4 | ....do .... | 3.15 | 3.06 | 0.47 | 0.79 |
| 695 | ♀ | ......do .............. | Sept. 3 | ....do .... | 3.14 | 3.15 | 0.43 | 0.75 |
| 983 | ♀ | Mountains, Southwest New Mexico. | Nov. 5 | ....do .... | 3.16 | 3.12 | 0.39 | 0.79 |
| 984 | ♂ | ......do .............. | Nov. 5 | ....do .... | 3.22 | 3.19 | 0.43 | 0.74 |
| 89 | ♀ | White Mountains, N. Mex. | Sept. 1 | Newberry. | 3.07 | 2.96 | 0.46 | 0.74 |

**69. *Poospiza bilineata*, (Cass.)—Black-throated Sparrow.**

Quite a common species in the vicinity of Wingate, N. Mex., in July, when the young in nesting-plumage were taken. Present in small numbers at Apache, Ariz., and abundant along the Gila the middle of September, where it habitually frequented the mezquite thickets. Is pre-eminently a bush and tree loving species.

| No. | Sex. | Locality. | Date. | Collector. | Wing. | Tail. | Bill. | Tarsus. |
|---|---|---|---|---|---|---|---|---|
| 432 | ♂ jun. | Wingate, N. Mex...... | July 14 | Henshaw. | 2.45 | 2.40 | 0.40 | 0.69 |
| 511 | ♀ jun. | Cave Spring, Ariz..... | Aug. 1 | ....do .... | 2.45 | 2.37 | 0.45 | 0.70 |
| 512 | ♀ jun. | ......do .............. | Aug. 1 | ....do .... | 2.47 | 2.51 | 0.42 | 0.72 |
| 611 | ♂ ad. | Mount Graham, Ariz.. | Sept. 21 | ....do .... | 2.50 | 2.59 | 0.40 | 0.74 |
| 675 | ♂ ad. | Apache, Ariz......... | Sept. 21 | ....do .... | 2.56 | 2.63 | 0.41 | 0.69 |
| 692 | ♂ | San Pedro, Ariz....... | Oct. 2 | ....do .... | 2.58 | 2.63 | 0.40 | 0.72 |
| 771 | ♂ ad. | Gila River, Ariz ...... | Oct. 2 | ....do .... | 2.50 | 2.66 | 0.43 | 0.71 |

**70. *Poospiza belli*, (Cass.), var. *nevadensis*, Ridgw.—Artemisia-Sparrow.**

Very abundant in the valleys of the San Pedro and Gila Rivers,

Arizona, in October. A well-marked race, differing from the true *belli* in larger size, as also in the paler coloration. Streaks on the back distinct and always present. These are usually entirely wanting in the typical *belli*, which is restricted in its range to California. Comparative measurements of the two races are appended.

*Poospiza belli.*

[In Smithsonian collection.]

| No. | Locality. | Wing. | Tail. | Bill. | Tarsus. |
|---|---|---|---|---|---|
| 6338 | Cosumnes River ........................... | 2.74 | 2.90 | 0.43 | 0.80 |
| 63652 | Saticoy, Cal ........................ | 2.75 | 3.00 | 0.40 | 0.75 |
| | California................................ | 2.43 | 2.31 | 0.32 | 0.81 |

*Poospiza, var. nevadensis.*

| No. | Sex. | Locality. | Date. | Collector. | Wing. | Tail. | Bill. | Tarsus. |
|---|---|---|---|---|---|---|---|---|
| 916 | ♂ | Gila River, Ariz ...... | Oct. 16 | Henshaw. | 3.02 | 3.02 | 0.43 | 0.83 |
| 917 | ♀ | ......do ............... | Oct. 16 | ....do .... | 2.95 | 2.85 | 0.43 | 0.77 |
| 921 | ♀ | ......do ............... | Oct. 16 | ....do .... | 3.10 | 2.97 | 0.42 | 0.86 |

71. *Spizella monticola*, (Gmel.)—Tree-Sparrow.

Not met with till at El Paso County, Colorado, where, according to Mr. Aiken, it is an abundant winter-resident. Found on the Colorado Chiquito, New Mexico, by Dr. Kennerly, in December. Specimens taken in this region average considerably lighter than in the east.

72. *Spizella socialis*, (Wils.), var. *arizonæ*, Cs.—Arizona Chipping-Sparrow.

Common through Arizona and New Mexico. Habits and notes appear identical with those of the eastern *socialis*. Two broods are raised in a season. A nest, found July 24, and containing young just hatched, was placed in a small piñon-tree a few feet from the ground.

| No. | Sex. | Locality. | Date. | Collector. | Wing. | Tail. | Bill. | Tarsus. |
|---|---|---|---|---|---|---|---|---|
| 449 | ♂ jun. | Wingate, N. Mex...... | July 16 | Henshaw . | 2.80 | 2.65 | 0.36 | 0.65 |
| 466 | ♂ jun. | Inscription Rock, N. Mex. | July 23 | ....do .... | 2.84 | 2.68 | 0.38 | 0.65 |
| 606 | ♂ jun. | Apache, Ariz.......... | Aug. 28 | ....do .... | 2.97 | 2.77 | 0.38 | 0.68 |
| 630 | Jun. | ......do ............... | Aug. 27 | ....do .... | 2.79 | 2.61 | 0.39 | 0.63 |
| 631 | ♂ jun. | ......do ............... | Aug. 6 | ....do .... | 2.79 | 2.60 | 0.40 | 0.67 |
| 621 | ♂ jun. | ......do ............... | Aug. 26 | ....do .... | 2.85 | 2.68 | 0.41 | 0.68 |
| 668 | ♂ jun. | ......do ............... | ......... | ....do .... | 2.86 | 2.65 | 0.39 | 0.66 |
| 723 | ♂ ad. | South of Apache, Ariz. | Sept. 1 | ....do .... | 2.76 | 2.60 | 0.40 | 0.65 |
| 847 | ♂ | Camp Grant, Ariz..... | Sept. 8 | ....do .... | 2.90 | 2.75 | 0.32 | 0.65 |
| 848 | ♀ | ......do ............... | Sept. 24 | ....do .... | 2.75 | 2.55 | 0.35 | 0.68 |
| 110 | ♀ ad. | Camp Bowie.......... | Sept. 9 | Newberry. | 2.75 | 2.60 | 0.38 | 0.65 |

73. *Spizella pallida* (Sw.), var. *breweri*, Cass.—Brewer's Sparrow.

Common, and generally distributed in Eastern Arizona. By the middle of August they had gathered into flocks, and, in company with other sparrows, were commonly seen seeking among weeds and bushes for seeds.

| No. | Sex. | Locality. | Date. | Collector. | Wing. | Tail. | Bill. | Tarsus. |
|-----|------|-----------|-------|------------|-------|-------|-------|---------|
| 631 | ♀ | Apache, Ariz.......... | Aug. 16 | Henshaw. | 2.45 | 2.55 | 0.31 | 0.65 |
| 632 | ♂ jun. | ......do .............. | Aug. 26 | ....do .... | 2.27 | 2.56 | 0.35 | 0.69 |
| 653 | ♀ | ......do .............. | Aug. 27 | ....do .... | 2.38 | 2.57 | 0.33 | 0.64 |
| 770 | ♂ | Gila River, Ariz....... | Sept. 1 | ....do .... | 2.40 | 2.62 | 0.35 | 0.71 |

74. *Melospiza melodia* (Wils.), var. *fallax*, Bd.—Western Song-Sparrow.
Not very common. Found in the White Mountains, Arizona, in August. Present, also, in small numbers, along the Gila.

| No. | Sex. | Locality. | Date. | Collector. | Wing. | Tail. | Bill. | Tarsus. |
|-----|------|-----------|-------|------------|-------|-------|-------|---------|
| 149 | ♀ | Bayard, N. Mex....... | Oct. 19 | Henshaw. | 2.57 | 2.95 | 0.46 | 0.82 |

75. *Melospiza lincolni*, Aud.—Lincoln's Finch.
Among the hordes of sparrows found along the Gila River, the middle of September, no one species compared at all in its abundance to this finch. The tall weeds and undergrowth were literally alive with these birds, dozens of which would be scared up at every step and alight on the neighboring trees. They spend all their time on the ground, searching for the small seeds and insects which constitute their food. When undisturbed, they are perfectly silent, but occasionally, when startled, emit a sharp chirp.

| No. | Sex. | Locality. | Date. | Wing. | Tail. | Bill. | Tarsus. |
|-----|------|-----------|-------|-------|-------|-------|---------|
| 746 | ♀ | Gila River, Ariz ................... | Sept. 11 | 2.35 | 2.40 | 0.48 | 0.76 |
| 785 | ♂ | ......do ...................... | Sept. 15 | 2.54 | 2.67 | 0.45 | 0.77 |
| 787 | ♀ | ......do ...................... | Sept. 15 | 2.57 | 2.70 | 0.43 | 0.75 |
| 789 | ♂ | ......do ...................... | Sept. 15 | 2.37 | 2.47 | 0.43 | 0.75 |
| 790 | ♀ | ......do ...................... | Sept. 15 | 2.47 | 2.65 | 0.48 | 0.84 |

76. *Peucæa ruficeps* (Cass.), var. *boucardi*, Sclat.
Under this variety of the Rufous-crowned Sparrow are included a series of ten sparrows, collected in Arizona from Apache southward and near Camp Bayard, N. Mex. From the typical *ruficeps*, as shown by specimens in the Smithsonian Institution, they differ in the generally darker coloration, especially shown in the rufous of the head, and in the stouter, darker bill, showing in these respects their relationship with *boucardi*.

Young birds in the nesting-plumage have the entire upper parts ashy-brown; beneath pale yellowish-white, profusely streaked across the breast and along the sides with dark-brown; greater wing-coverts tipped with fulvous; secondaries margined outwardly with dull-rufous.

This sparrow was found to prefer rocky localities, generally in the close vicinity of the streams. In some places it was not uncommon, usually in small companies of from three to eight. I never saw it near the pines, and, at this season at least, doubt it ever being found among them. Indeed, all its habits and motions, as it busies itself searching for food among the rocks and bushes, are exceedingly similar to the Song-Sparrow (*M. fallax*), for which I mistook it more than once; its chirp of alarm was very similar.

Bill dark-brown above, paler below; legs and feet light-brown.

| No. | Sex. | Locality. | Date. | Collector. | Wing. | Tail. | Bill. | Tarsus. |
|-----|------|-----------|-------|------------|-------|-------|-------|---------|
| 591 | ♂ ad. | Apache, Ariz........ ....... | Aug. 21 | Henshaw . | 2.53 | 2.73 | 0.50 | 0.75 |
| 713 | ♀ ad. | Thirty miles south of Apache, Ariz. | Sept. 7 | .... do .... | 2.47 | 2.76 | 0.51 | 0.75 |
| 719 | ♂ ad. | ....do ................. | Sept. 8 | .... do .... | 2.38 | 2.86 | 0.47 | 0.78 |
| 720 | ♀ jun. | ....do ................. | Sept. 8 | .... do .... | 2.38 | 2.63 | 0.47 | 0.80 |
| 744 | ♀ ad. | ....do ................. | Sept. — | .... do .... | 2.35 | 2.72 | 0.50 | 0.79 |
| 758 | ♂ | Gila River, Ariz ...... | Sept. 11 | ............ | 2.30 | 2.73 | 0.51 | 0.74 |
| 764 | ♂ ad. | ....do ...... ......... | Sept. 12 | Turner ... | 2.55 | 2.92 | 0.50 | 0.77 |
| 890 | ♂ | Camp Grant, Ariz..... | Sept. 27 | Henshaw . | 2.57 | 3.20 | 0.46 | 0.81 |
| 898 | ♂ jun. | ....do ................. | Sept. 30 | .... do .... | 2.44 | 3.76 | 0.45 | 0.80 |
| 946 | ♂ | Bayard, N. Mex........ | Oct. 19 | .... do .... | 2.63 | 3.15 | 0.47 | 0.74 |

77. *Peucæa*, sp. (?)

Feathers above with dark-brown centers, and edged conspicuously with fulvous; brightest on the rump, where each feather is broadly tipped with the same; beneath pale ochraceous-yellow, becoming strong fulvous on the flanks and under tail-coverts; upper parts of breast and throat strongly and sides less distinctly marked with longitudinal streaks of black; wing-coverts edged and tipped with strong fulvous; inner secondaries bordered with same, but darker; tail-feathers black, margined with dull-rufous; bend of wing edged with light-yellow; bill above dark-brown, paler beneath.

| No. | Sex. | Locality. | Date. | Collector. | Wing. | Tail. | Bill. | Tarsus. |
|-----|------|-----------|-------|------------|-------|-------|-------|---------|
| 878 | ♂ jun. | Camp Grant, Ariz..... . | Sept. 27 | Henshaw . | 2.50 | 2.58 | 0.47 | 0.80 |

The above is a description of a sparrow taken at Camp Grant. It was started from the long grass on the open plain, and no others were seen. From a comparison made with specimens in the Smithsonian Institution, it appears not to be the young of any species of the genus known to inhabit the United States, but may perhaps be one of several Mexican forms, of which the Institution has no examples. Its immature condition renders it exceedingly difficult to arrive at any satisfactory conclusion respecting it, and I therefore deem it best to leave its determination till better specimens or a larger series are at hand for comparison.

78. *Passerella townsendii* (Aud.), var. *schistacea*, Bd.

Probably very rare in this region, as a single specimen, taken in a small cañon south of Apache, was the only one seen during the season.

| No. | Sex. | Locality. | Date. | Collector. | Wing. | Tail. | Bill. | Tarsus. |
|-----|------|-----------|-------|------------|-------|-------|-------|---------|
| 738 | ♀ jun. | Forty miles south of Apache, Ariz. | Sept. 1 | Henshaw . | 3.00 | 3.45 | 0.48 | . 0.86 |

79. *Calamospiza bicolor*, (Towns.)—Lark-Bunting; White-winged Black-bird.

A few in the worn breeding-plumage were seen in the neighborhood of Zuni, N. Mex., in July. Leaving here, the species was not again met with till October, when they were found in large flocks in the San Pedro and Gila Valleys, Arizona. They feed almost entirely at this season upon the seeds of various grasses, and, when engaged in searching for these, show little of the shyness attributed to them at other periods of the year. By the middle of October, the males have assumed the plumage of the females, and are indistinguishable from them and the young, except that the streakings underneath are heavier and blacker, particularly about the throat, and there is also much black on the wings.

| No. | Sex. | Locality. | Date. | Collector. | Wing. | Tail. | Bill. | Tarsus. |
|-----|------|-----------|-------|------------|-------|-------|-------|---------|
| 503 | ♂ ad. | Zuni, N. Mex .......... | July 25 | Henshaw . | 3.45 | 2.92 | 0.56 | 0.93 |
| 899 | ♀ | San Pedro, Ariz........ | Oct. 3 | Magnet... | 3.25 | 2.74 | 0.52 | 0.91 |
| 907 | ♀ | ....do .............. | Oct. 3 | .... do .... | 3.28 | 2.65 | 0.54 | 0.90 |
| 913 | ♀ | ....do .............. | Oct. 3 | Henshaw . | 3.25 | 2.54 | 0.52 | 0.90 |
| 930 | ♂ ad. | Gila River, Ariz ...... | Oct. 17 | .... do .... | 3.55 | 2.95 | 0.60 | 0.95 |
| 931 | ♂ ad. | ....do .............. | Oct. 17 | .... do .... | 3.40 | 3.03 | 0.53 | 0.90 |
| 935 | ♂ ad. | ....do .............. | Oct. 17 | .... do .... | 3.41 | 2.85 | 0.53 | 0.90 |
| 936 | ♂ | ....do .............. | Oct. 17 | .... do .... | 3.41 | 2.95 | 0.52 | 0.93 |
| 937 | ♀ | ....do .............. | Oct. 17 | .... do .... | 3.37 | 2.87 | 0.56 | 0.95 |
| 940 | ♀ | ....do .............. | Oct. 17 | .... do .... | 3.00 | 2.67 | 0.54 | 0.90 |

80. *Euspiza americana*, (Gm.)—Black-throated Bunting.

A specimen taken near San Pedro, Ariz., September 3. This was apparently a mere straggler, as the species was nowhere else met with.

| No. | Sex. | Locality. | Date. | Collector. | Wing. | Tail. | Bill. | Tarsus. |
|-----|------|-----------|-------|------------|-------|-------|-------|---------|
| 896 | ♀ jun. | San Pedro, Ariz........ | Sept. 3 | Henshaw . | 3.18 | 2.50 | 0.55 | 0.86 |

81. *Hedymeles melanocephalus*, (Sw.)—Black-headed Grossbeak.

Rather numerous in the vicinity of Wingate, N. Mex., where it breeds. Common also at Apache, keeping in the tall cottonwoods and thickets of the streams.

| No. | Sex. | Locality. | Date. | Collector. | Wing. | Tail. | Bill. | Tarsus. |
|-----|------|-----------|-------|------------|-------|-------|-------|---------|
| 95 | ♀ ad. | Apache, Ariz ......... | Sept. 12 | Newberry . | 3.19 | 3.15 | 0.73 | 0.86 |
| 642 | ♂ jun. | ....do .............. | Aug. 28 | Henshaw . | 3.73 | 3.12 | 0.76 | 0.85 |
| 690 | ♀ jun. | ....do .............. | Sept. 3 | .... do .... | 4.06 | 3.40 | 0.70 | 0.88 |

82. *Guiraca cœrulea*, (L.)—Blue Grossbeak.

Among the bushes and copses that line the river-bank at Apache this species was not uncommon toward the end of August, and I doubt not that they breed here. They were very shy, and an occasional glimpse of a bird as it flitted from one clump to another was all that could be obtained.

| No. | Sex. | Locality. | Date. | Collector. | Wing. | Tail. | Bill. | Tarsus. |
|-----|------|-----------|-------|-----------|-------|-------|-------|---------|
| 629 | ♂ jun. | Apache, Ariz .......... | Aug. 27 | Henshaw. | 3. 43 | 2. 90 | 0. 42 | 0. 75 |

83. *Cyanospiza amœna*, (Say.)—Lazuli Finch.

Common at Apache, where it frequented the bushes and weeds in search of seeds. In a male taken in October the blue is clouded and almost obscured by rufous, which overspreads the whole plumage.

| No. | Sex. | Locality. | Date. | Collector. | Wing. | Tail. | Bill. | Tarsus. |
|-----|------|-----------|-------|-----------|-------|-------|-------|---------|
| 598 | ♀ jun. | Apache, Ariz .......... | Aug. 23 | Henshaw. | 2. 46 | 2. 21 | 0. 37 | 0. 63 |
| 609 | ♂ ad. | ....do ................ | Aug. 24 | .... do .... | 2. 87 | 2. 41 | 0. 39 | 0. 67 |
| 623 | ♀ juu. | ....do ................ | Aug. 26 | .... do .... | 2. 68 | 2. 32 | 0. 38 | 0. 65 |
| 707 | ♂ jun. | ....do ................ | Sept. — | .... do .... | 2. 85 | 2. 25 | 0. 40 | 0. 66 |
| 891 | ♂ ad. | San Pedro, Ariz....... | Oct. 2 | .... do .... | 2. 85 | 2. 52 | 0. 40 | 0. 63 |

84. *Pipilo maculatus*, Sw., var. *megalonyx* Bd.—Long-spurred Towhee.

Apparently the only Black Pipilo inhabiting this region. Though not very common, it was found everywhere along the route in Eastern Arizona and in New Mexico till late in October.

| No. | Sex. | Locality. | Date. | Collector. | Wing. | Tail. | Bill. | Tarsus. |
|-----|------|-----------|-------|-----------|-------|-------|-------|---------|
| 448 | ♂ ad. | Wingate, N. Mex ..... | July 16 | Henshaw. | 3. 36 | 4. 02 | 0. 58 | 1. 05 |

85. *Pipilo fuscus* Sw., var. *mesoleucus*, Bd.—Cañon-Finch.

I did not detect this species on the Gila, where the Abert's Finch was very numerous. When nearing Camp Grant, my attention was attracted by hearing notes issuing from a thicket on the sides of a rocky cañon, which I was confident I had never before heard, and a short search soon revealed the author to be this finch. The notes are much deeper and harsher than those of the Abert's Finch. The habits of the two birds appear much the same, but the present bird seems rather to prefer rocky cañons to more open situations.

| No. | Sex. | Locality. | Date. | Collector. | Wing. | Tail. | Bill. | Tarsus. |
|-----|------|-----------|-------|-----------|-------|-------|-------|---------|
| 866 | ♂ | Camp Grant, Ariz..... | Sept. 24 | Henshaw . | 3. 64 | ...... | 0. 60 | 0. 98 |
| 126 | ...... | Bowie, Ariz........... | Oct. 10 | Newberry. | 3. 58 | 4. 24 | 0. 57 | 0. 98 |
| 4a | ...... | San Carlos, Ariz ...... | Sept. 13 | Maguet .. | 3. 57 | 4. 55 | 0. 58 | 0. 99 |

86. *Pipilo aberti*, Bd.—Abert's Towhee.

This was a very abundant species along the Gila River, which was the only point where it was seen. It frequented the thickest brush, whence its loud, peculiar chirp could be heard issuing at all times. It was gregarious at this time, considerable numbers being found together, and always showed great shyness, betaking itself on the least alarm to the impenetrable mezquite-thickets. All were molting.

| No. | Sex. | Locality. | Date. | Collector. | Wing. | Tail. | Bill. | Tarsus. |
|---|---|---|---|---|---|---|---|---|
| 766 | ♂ | Gila River, Ariz ...... | Sept. 14 | Henshaw. | 3.70 | 4.92 | 0.58 | 1.08 |
| 767 | ♀ | ......do ............. | Sept. 14 | ....do .... | 3.55 | 4.71 | 0.60 | 1.09 |
| 779 | ♂ | ......do ............. | Sept. 15 | ....do .... | 3.46 | 4.81 | 0.60 | 1.07 |
| 796 | ♀ jun. | ......do ............. | Sept. 14 | ....do .... | 3.53 | 4.68 | 0.60 | 1.11 |
| 797 | ♂ jun. | ......do ............. | Sept. 16 | ....do .... | 3.41 | 4.45 | 0.60 | 1.05 |
| 798 | ♂ jun. | ......do ............. | Sept. 16 | ....do .... | 3.61 | 4.75 | 0.60 | 1.09 |
| .... | | | | | 3.55 | 4.75 | 0.58 | 1.00 |

87. *Pipilo chlorurus*, (Towns.)—Green-tailed Finch.

Rather numerous on the brushy streams near Apache, Ariz., in August. Generally distributed during the fall-migration.

| No. | Sex. | Locality. | Date. | Collector. | Wing. | Tail. | Bill. | Tarsus. |
|---|---|---|---|---|---|---|---|---|
| 725 | ♀ jun. | Near Apache, Ariz .... | Sept. 8 | Henshaw. | 2.87 | 3.40 | 0.48 | 0.93 |
| 760 | ♀ jun. | South of Apache, Ariz. | Sept. 11 | ....do .... | 2.87 | 3.55 | 0.52 | 0.90 |
| 5 | Jun. | San Carlos, Ariz ...... | Sept. 13 | Magnet .. | 2.83 | 3.39 | 0.50 | 0.94 |

ALAUDIDÆ (the Larks).

88. *Eremophila alpestris* (Forst.), var. *chrysolœma*, Wagl.

The young were taken near Wingate, N. Mex., June 30, by Dr. Newberry, jr. After September, the species was found gathered in large flocks and scattered over the dry and arid plains, where they feed upon the seeds and insects which they pick up among the sage-brush and bushes. Later, in the latter part of November, the plains between Wingate and Santa Fé were fairly alive with these birds, and flocks numbering thousands were met with at short intervals.

ICTERIDÆ (the Orioles).

89. *Agelaius phœniceus*, (L.)—Red-winged Blackbird.

A few noticed at Apache the latter part of August, associating with flocks of the succeeding species.

90. *Xanthocephalus icterocephalus*, (Bon.)—Yellow-headed Blackbird.

Not common near Zuni, N. Mex., the latter part of July. Present at Apache in considerable flocks in the marshy spots along the river.

| No. | Sex. | Locality. | Date. | Collector | Wing. | Tail. | Bill. | Tarsus. |
|---|---|---|---|---|---|---|---|---|
| 628 | ♂ jun. | Apache, Ariz.......... | Aug. 27 | Henshaw. | 5.54 | 4.13 | 0.80 | 1.31 |

91. *Scolecophagus cyanocephalus*, (Wagl.)—Brewer's Blackbird.

A generally well-distributed species, both in New Mexico and Arizona. Unlike its eastern congener, the Rusty Blackbird, which is pre-eminently a swamp and marsh loving species, the Brewer's seems little inclined to prefer such localities, but is often found, especially in the fall, on the outskirts of settlements, haunting the corrals and barn-yards.

CORVIDÆ (the Crows).

92. *Corvus corax* L., var. *carnivorus*, Bartr.—American Raven.

An ever-present species throughout our route. In the fall and winter they congregate just outside the settlements, frequenting the corn-fields and the ranges, where horses and cattle are pastured. In the wilderness they were most often met with in pairs, and here seemed shy and suspicious. Their sharp eyes were quick to spy out our presence, and always in early morning the croakings of one or two pairs were heard from the neighboring trees or rocks, where they were impatiently awaiting till our departure should enable them to swoop down into the camp, and quarrel over any stray morsels left behind.

93. *Corvus cryptoleucus*, Couch.—White-necked Crow.

I did not detect the presence of this bird in either New Mexico or Arizona. A large flock was seen near Colorado Springs in December, in which region the observations of Mr. Aiken show it to be a very abundant species.

94. *Picicorvus columbianus*, (Wils.)—Clarke's Crow.

About the middle of August this species was apparently not uncommon in the White Mountains, Arizona, where a specimen was obtained by Dr. Newberry, jr. At this season it was very restless and shy.

| No. | Sex. | Locality. | Date. | Collector. | Wing. | Tail. | Bill. | Tarsus. |
|---|---|---|---|---|---|---|---|---|
| 54 | ♂ ad. | White Mts., Ariz...... | Aug. 20 | Newberry. | 7. 38 | 4. 60 | 1. 57 | 1. 43 |

95. *Gymnokitta cyanocephala*, Pr. Max.—Maximilian's Jay.

A large flock of these jays were seen near Silver City, N. Mex., October, busily engaged on the ground feeding upon grass-seeds. Those in the rear kept flying up and alighting in the front rank, the whole flock thus keeping in continual motion. Near Tulerosa late in November, I found the species an abundant one, and chiefly frequenting the pinicoline trees. Their habits here, however, seemed to imply a scarcity of their favorite food, which are the various seeds of the coniferous trees, for I saw a large flock engaged in catching insects on the wing, and in this novel occupation they displayed no little dexterity. From the tops of the pine-trees they ascended to a considerable height, when, hovering for an instant, they would snap up an insect and return to near the former position, remain for a moment, and again make an essay.

| No. | Sex. | Locality. | Date. | Collector. | Wing. | Tail. | Bill. | Tarsus. |
|---|---|---|---|---|---|---|---|---|
| 32 | ♀ jun. | Wingate, N. Mex...... | July 16 | Newberry. | 5. 80 | 4. 46 | 1. 28 | 1. 40 |

96. *Pica melanoleuca* Vieill., var. *hudsonica*, Sab.—Magpie.

A single bird in nesting-plumage was shot by Dr. Newberry, jr., at the Rio Puerco, sixty miles west of Wingate, N. Mex. Farther south than this the species was not met with; and if occurring in Eastern and Southeastern Arizona, it must, I think, be rare.

97. *Cyanura stelleri* (Gm.), var. *macrolopha*, Bd.—Long-crested Jay.

This jay is one of the most characteristic birds of the western woods, conspicuous alike for its beautiful plumage and its loud and peculiar notes. In habits it is very largely, though not exclusively, pinicoline, being generally found throughout the heavy pine-timber of the mountainous districts. It was observed by us to be numerous in such localities, both in Arizona and New Mexico. Like most others of the family, it is gifted with considerable curiosity, which, however, is rarely sufficient to overcome its naturally rather suspicious disposition. During the fall they usually move about in small parties of six or eight, and seem to spend considerable time on the ground, hunting after seeds, acorns, and berries, which supplement at this season their usual fare, consisting of the seeds of coniferous trees. I have often come suddenly upon a party when thus silently and busily engaged, searching among the bushes, often not less to my own than to their surprise. A single note was sufficient to alarm the whole flock, when they would betake themselves to the nearest tree, and watch my every motion with evident interest, all the while keeping up a constant chattering and screaming. Their natural distrust, however, would soon induce them to place a wider interval between us, and to approach a second time when they had once flown would have been no easy matter.

| No. | Sex. | Locality. | Date. | Collector. | Wing. | Tail. | Bill. | Tarsus. |
|---|---|---|---|---|---|---|---|---|
| 689 | ♂ | Apache Mts., Ariz..... | Sept. 1 | Henshaw. | 5.73 | 5.41 | 1.17 | 1.55 |
| 753 | ♀ jun. | South of Apache, Ariz. | Sept. 11 | ....do .... | 5.51 | 5.37 | 1.12 | 1.65 |
| 4 | ...... | ......do ............... | Sept. 11 | Magnet .. | 5.64 | 5.40 | 1.16 | 1.72 |
| 940 | ♂ | Gila River, Ariz ...... | Oct. 17 | Henshaw. | 5.93 | 5.43 | 1.17 | 1.68 |

98. *Cyanocitta floridana* (Bartr.), var. *woodhousii*, Bd.—Woodhouse's Jay.

A very common species at Wingate, N. Mex., Apache and Grant, Ariz., and elsewhere. Frequents particularly the shrubbery and thickets of the hill-sides. Subsists upon nuts, acorns, seeds, berries, insects, and is, in fact, almost omnivorous.

| No. | Sex. | Locality. | Date. | Collector. | Wing. | Tail. | Bill. | Tarsus. |
|---|---|---|---|---|---|---|---|---|
| 114 | ♀ | Arizona ............... | Sept. 28 | Newberry. | 5.04 | 5.68 | 1.11 | 1.40 |
| 589 | ♂ jun. | Apache, Ariz ......... | Sept. 21 | Henshaw . | 4.93 | 5.71 | 1.06 | 1.55 |
| 868 | ♂ | Camp Grant, Ariz..... | Sept. 24 | ....do .... | 5.35 | 5.98 | 1.15 | 1.60 |

99. *Perisoreus canadensis* (L.), var. *capitalis*, Bd.—Rocky Mountain Gray Jay.

Collected in the White Mountains, Arizona, by Dr. Newberry, jr., who found it not uncommon in the forests of spruce and pine.

| No. | Sex. | Locality. | Date. | Collector. | Wing. | Tail. | Bill. | Tarsus. |
|---|---|---|---|---|---|---|---|---|
| 78 | ♂ ad. | White Mountains, Ariz. | Aug. 27 | Newberry. | 6.23 | 5.98 | 0.90 | 1.35 |

100. *Cyanocitta ultramarina* (Bp.), var. *arizonæ*, Ridg.—Arizona Jay.

I first saw this species when encamped in a narrow, rocky cañon, thirty miles south of Apache, Ariz. The sides of the cañon and the neighboring heights were well covered with a small species of oak, which were habitually frequented by these birds, and the fruit of which doubtless forms a part of its food. They were not very numerous, but appeared to keep in small flocks of from six to twelve. Occasionally they were seen upon the ground, hunting for seeds, berries, and insects, but the species seems to be rather more arboreal in its habits than any others of the genus with which I am acquainted. Their notes are essentially garruline in character, but are surprisingly weak for the size of the bird, while it is far less noisy than others of the family. At Camp Grant they were rather more common, frequenting about the same localities. They were quite shy, showing little or no curiosity, but on discovering my presence would immediately make a hasty retreat through the trees, and it was only when thus disturbed that their cries were heard. In New Mexico, I observed the species as far north as Camp Bayard. Hitherto known but from two localities in New Mexico, viz, Fort Buchanan and the Copper Mines. In summer, its northward range is probably limited to about latitude 34°. An immature bird, just molting the nesting-plumage, has the blue of the upper parts mixed with dull ash. The bill is flesh-colored, the upper mandible flesh-colored at tip.

Bill of adult black; immature birds black, varied with flesh-color.

| No. | Sex. | Locality. | Date. | Collector. | Wing. | Tail. | Bill. | Tarsus. |
|---|---|---|---|---|---|---|---|---|
| 733 | ♀ jun. | Thirty miles south of Apache, Ariz. | Sept. 11 | Henshaw. | 6.30 | 5.91 | 1.12 | 1.68 |
| 734 | ♂ jun. | ......do ............... | Sept. 11 | ....do .... | 6.47 | 6.13 | 1.23 | 1.65 |
| 757 | ♂ jun. | ......do ............... | Sept. 12 | ....do .... | 6.72 | 6.31 | 1.16 | 1.62 |
| 845 | ♂ ad. | Camp Grant, Ariz..... | Sept. 24 | ....do .... | 6.82 | 6.44 | 1.25 | 1.56 |
| 896 | ♂ jun. | ......do ............... | Sept. 30 | ....do .... | 6.32 | 5.88 | 1.22 | 1.68 |
| 897 | ♂ jun. | ......do ............... | Sept. 30 | ....do .... | 6.44 | 5.92 | 1.25 | 1.63 |

TYRANNIDÆ (the Flycatchers).

101. *Tyrannus verticalis*, Say.—Arkansas Flycatcher.

Not numerous at Wingate, N. Mex. Present in small numbers at Apache, Ariz. Keeps in the open-wooded districts.

| No. | Sex. | Locality. | Date. | Collector. | Wing. | Tail. | Bill. | Tarsus. |
|---|---|---|---|---|---|---|---|---|
| 702 | ♀ jun. | Apache, Ariz.......... | Sept. 6 | Henshaw. | 4.55 | 3.68 | 0.76 | 0.76 |
| 708 | ♂ jun. | ......do .............. | Sept. 7 | ....do .... | 4.95 | 3.85 | 0.78 | 0.70 |

102. *Tyrannus vociferans*, Sw.—Cassin's Flycatcher.

A rather common species among the sage-brush about Fort Wingate. Frequents also the open, brushy ravines, and altogether seemed to be less of a tree-loving species than the preceding. Noted, also, at various points in Eastern Arizona to Fort Bowie.

| No. | Sex. | Locality. | Date. | Collector. | Wing. | Tail. | Bill. | Tarsus. |
|---|---|---|---|---|---|---|---|---|
| 457 | ♂ ad. | Neutria, N. Mex........ | July 19 | Henshaw . | 5.25 | 4.01 | 0.80 | 0.71 |
| 2d | ♂ ad. | Wingate, N. Mex...... | July 15 | Newberry. | 5.25 | 3.91 | 0.86 | 0.78 |
| 465 | ♀ jun. | Inscription Rock, N. Mex. | July 23 | Henshaw . | 5.13 | 3.95 | 0.75 | 0.75 |
| 497 | ♀ ad. | .......do ............. | July 24 | ....do .... | 5.16 | 3.90 | 0.91 | 0.78 |
| 609 | ♀ | Mount Turnbull, Ariz . | Sept.21 | ....do .... | 5.15 | 3.95 | 0.83 | 0.75 |
| 8a | ♂ | ......do ............. | Sept.20 | Magnet... | 5.14 | 3.85 | 0.87 | 0.78 |
| 778 | ♀ jun. | Gila River, Ariz ...... | Sept.15 | Henshaw . | 4.93 | 3.96 | 0.79 | 0.76 |
| 793 | ♀ ad. | .......do ............. | Sept.16 | ....do .... | 4.92 | 3.81 | 0.79 | 0.75 |

103. *Myiarchus crinitus* (L.), var. *cinerascens*, Lawr.—Ash-throated Fly-catcher.

Less abundant than the preceding, but inhabiting much the same region. In choice of localities, it evinced a similar taste, as it affected the open plains and creek-bottoms grown up to brush rather than the more densely-wooded districts.

| No. | Sex. | Locality. | Date. | Collector. | Wing. | Tail. | Bill. | Tarsus. |
|---|---|---|---|---|---|---|---|---|
| 33 | ♀ ad. | Wingate, N. Mex...... | July 16 | Newberry. | 3.73 | 3.62 | 0.73 | 0.89 |
| 447 | ♀ ad. | .......do ............. | July 15 | Henshaw . | 3.75 | 3.69 | 0.77 | 0.85 |
| 475 | ♀ jun. | Inscription Rock, N. Mex. | July 23 | ....do .... | 3.67 | 3.64 | 0.70 | 0.87 |
| 438 | ♂ ad. | Wingate, N. Mex...... | July 15 | ....do .... | 4.00 | 3.94 | 0.75 | 0.88 |

104. *Sayornis nigricans*, (Sw.)—Black Flycatcher.

Present throughout Eastern Arizona, where it is a common inhabitant of the brush-lined streams, and is to be seen constantly in pursuit of flying insects. Its habits seem to correspond pretty closely with those of the eastern pewee (*S. fuscus*).

| No. | Sex. | Locality. | Date. | Collector. | Wing. | Tail. | Bill. | Tarsus. |
|---|---|---|---|---|---|---|---|---|
| 118 | ♀ jun. | Bowie, Ariz........... | Sept. 6 | Newberry. | 3.25 | 3.25 | 0.58 | 0.66 |
| 522 | ♂ jun. | Apache, Ariz........... | Aug. 5 | Henshaw . | 3.28 | 2.86 | 0.61 | 0.65 |
| 844 | ♀ | Grant, Ariz .......... | Sept. 24 | ....do .... | 3.46 | 3.36 | 0.57 | 0.70 |
| 855 | ♀ | ......do ............. | Sept. 24 | ....do .... | 3.32 | 2.07 | 0.57 | 0.68 |

105. *Sayornis sayus*, (Bon.)—Say's Flycatcher.

In the neighborhood of Wingate, in July, both the old and young of this flycatcher were abundant. For the most part, they were found inhabiting the open sage-brush, or the open and rocky hill-sides scantily clothed with brush and a few scattering piñon-trees.

| No. | Sex. | Locality. | Date. | Collector. | Wing. | Tail. | Bill. | Tarsus. |
|---|---|---|---|---|---|---|---|---|
| 1 | ♂ ad. | Santa Fé, N. Mex | June 10 | Newberry. | 3.72 | 3.32 | 0.60 | 0.42 |
| 440 | ♂ jun. | Wingate, N. Mex | July 15 | Henshaw. | 4.24 | 3.52 | 0.68 | 0.80 |
| 441 | ♀ jun. | ......do | July 15 | ....do .... | 3.97 | 3.28 | 0.62 | 0.75 |
| 442 | ♀ jun. | ......do | July 15 | ....do .... | 3.95 | 3.15 | 0.65 | 0.73 |
| 794 | ♂ ad. | Gila River, Ariz | Sept. 17 | ....do .... | 4.17 | 3.45 | 0.59 | 0.77 |

106. *Contopus borealis*, (Sw.)—Olive-sided Flycatcher.

Common at Apache, Ariz, in August, keeping chiefly in the neighborhood of the river. Also noticed further south on the Gila.

Upper mandible black, lower light-brown; legs and feet black.

| No. | Sex. | Locality. | Date. | Collector. | Wing. | Tail. | Bill. | Tarsus. |
|---|---|---|---|---|---|---|---|---|
| 584 | ♀ ad. | Apache, Ariz | Aug. 2 | Henshaw. | 4.15 | 2.93 | 0.70 | 0.60 |
| 597 | ♂ ad. | ......do | Aug. 23 | ....do .... | 4.30 | 2.18 | 0.75 | 0.64 |

107. *Contopus pertinax*, Cab.—Coues's Flycatcher; Mexican Olive-sided Flycatcher.

Apparently a very rare species. I met with it but on a single occasion, in the heavy pine-woods near Apache, Ariz. While riding along, I was attracted by certain loud, harsh, screaming notes, and, dismounting, after much trouble and dodging among the trees, I succeeded in getting a sight at the authors, and found that a pair of old birds were feeding several young, the latter being fully fledged, and not distinguishable in colors at the distance from the old. As a result of two shots I obtained both the old birds, and found them to be this species. The plumage was very much worn and bleached.

Iris brown; bill above black, below bright-yellow; legs and feet black.

| No. | Sex. | Locality. | Date. | Collector. | Wing. | Tail. | Bill. | Tarsus. |
|---|---|---|---|---|---|---|---|---|
| 549 | ♂ ad. | White Mountains, Ariz. | Aug. 10 | Henshaw. | 4.30 | 3.72 | 0.75 | 0.65 |
| 550 | ♀ ad. | ......do | Aug. 10 | .... do .... | 3.80 | 3.25 | 0.75 | 0.65 |

108. *Contopus virens* (L.), var. *richardsonii*, Sw.—Western Wood-Pewee.

An especially abundant species at Inscription Rock, N. Mex., where both old and young were seen July 23 among the cedars. Common along our route in Eastern Arizona. Arboreal in its habits at all seasons.

Iris brown; bill black; lower mandible yellow; tip brown; feet black.

| No. | Sex. | Locality. | Date. | Collector. | Wing. | Tail. | Bill. | Tarsus. |
|---|---|---|---|---|---|---|---|---|
| 583 | ♂ jun. | Apache, Ariz | Aug. 21 | Henshaw. | 3.40 | 2.70 | 0.49 | 0.49 |
| 594 | ♀ jun. | ......do | Aug. 22 | .... do .... | 3.27 | 2.61 | 0.47 | 0.51 |
| 714 | ♂ | ......do | Sept. 7 | .... do .... | 3.30 | 2.66 | 0.53 | 0.51 |
| 783 | Jun. | Gila River, Ariz | Sept. 15 | .... do .... | 3.09 | 2.63 | 0.52 | 0.52 |

109. *Empidonax pusillus*, (Sw.)—Little Flycatcher.

Seen occasionally on the streams about Apache in August.

Bill black above, pale-brown beneath.

| No. | Sex. | Locality. | Date. | Collector. | Wing. | Tail. | Bill. | Tarsus. |
|---|---|---|---|---|---|---|---|---|
| 560 | ♀ ad. | White Mountains, Ariz. | Aug. 11 | Henshaw . | 2.65 | 2.45 | 0.51 | 0.60 |
| 622 | ♂ jun. | Apache, Ariz ............ | Aug. 26 | .... do .... | 2.56 | 2.33 | 0.46 | 0.59 |
| 665 | ♂ jun. | Near Apache, Ariz ..... | Sept. 8 | .... do .... | 2.63 | 2.55 | 0.52 | 0.60 |
| 703 | ♀ jun. | ......do ............. | Sept. 8 | .... do .... | 2.60 | 2.38 | 0.52 | 0.67 |

110. *Empidonax flaviventris* Bd., var. *difficilis*, Bd.—Western Yellow-bellied Flycatcher.

As in summer, seems to prefer the narrow cañons and secluded localities. Common, and generally distributed.

| No. | Sex. | Locality. | Date. | Collector. | Wing. | Tail. | Bill. | Tarsus. |
|---|---|---|---|---|---|---|---|---|
| 454 | ♂ ad. | Wingate, N. Mex...... | July 18 | Henshaw . | 2.95 | 2.63 | 0.53 | 0.66 |
| 467 | ♂ jun. | Inscription Rock, N. Mex. | July 23 | .... do .... | 2.54 | 2.35 | 0.48 | 0.64 |
| 722 | ♀ | South Apache, N. Mex. | Sept. 8 | .... do .... | 2.60 | 2.40 | 0.53 | 0.71 |

111. *Empidonax obscurus*, (Sw.)—Wright's Flycatcher.

A common and in the fall a pretty generally-distributed species, frequenting not only the copses and thickets of the streams, but also the groves of oaks and other deciduous trees. Quick and energetic in its motions. In this respect in marked contrast to the succeeding species, which it resembles somewhat in its appearance.

| No. | Locality. | Date. | Collector. | Wing. | Tail. | Bill. | Tarsus. |
|---|---|---|---|---|---|---|---|
| 447 | Wingate, N. Mex ............ | July 15 | Henshaw . | 2.70 | 2.55 | 0.52 | 0.68 |
| 479 | Inscription Rock, N. Mex .... | July 23 | .... do .... | 2.80 | 2.58 | 0.50 | 0.65 |
| 518 | Apache, Ariz ................. | Aug. 14 | .... do .... | 2.70 | 2.60 | ...... | 0.70 |
| 666 | ......do ..................... | Sept. 1 | .... do .... | 2.55 | 2.54 | 0.47 | 0.65 |
| 696 | ......do ..................... | Sept. 4 | .... do .... | 2.80 | 2.55 | 0.50 | 0.68 |
| 721 | South of Apache, Ariz ....... | Sept. 10 | .... do .... | 2.60 | 2.46 | 0.50 | 0.70 |
| 726 | ......do ..................... | Sept. 10 | .... do .... | 2.85 | 2.60 | 0.53 | 0.70 |

112. *Empidonax hammondii*, De Vesey.—Hammond's Flycatcher.

Perhaps the most common of the *Empidonaces*. From Apache southward, seen all along the route, preferring generally the oak-groves, from the low limbs of which trees it sallied forth after insects. Has the same rather listless habits as noticed in the summer, with less dash and spirit than any of the small flycatchers.

| No. | Sex. | Locality. | Date. | Collector. | Wing. | Tail. | Bill. | Tarsus. |
|---|---|---|---|---|---|---|---|---|
| 681 | ♂ | Apache, Ariz .......... | Sept. 2 | Henshaw . | 2.54 | 2.33 | 0.40 | 0.60 |
| 710 | ♂ | ......do ............. | Sept. 7 | .... do .... | 2.75 | 2.52 | 0.42 | 0.64 |
| 711 | ♀ | ......do ............. | Sept. 7 | .... do .... | 2.61 | 2.56 | 0.42 | 0.62 |
| 729 | ♂ | ......do ............. | Sept. 10 | .... do .... | 2.64 | 2.63 | 0.42 | 0.69 |
| 728 | ♀ | ......do ............. | Sept. 10 | .... do .... | 2.64 | 2.43 | 0.40 | 0.60 |
| 730 | ♂ | Gila River, Ariz....... | Sept. 10 | .... do .... | 2.73 | 2.45 | 0.40 | 0.60 |
| 759 | ♀ | ......do ............. | Sept. 10 | .... do .... | 2.75 | 2.43 | 0.41 | 0.65 |
| 772 | ♂ | ......do ............. | Sept. 15 | .... do .... | 2.80 | 2.58 | 0.40 | 0.60 |
| 788 | ♀ | ......do ............. | Sept. 15 | .... do .... | 2.50 | 2.12 | 0.42 | 0.59 |
| 948 | ♀ | Bayard, N. Mex ....... | Sept. 19 | .... do .... | 2.69 | 2.48 | 0.40 | 0.61 |

113. *Mitrephorus fulvifrons* (Giraud), var. *pallescens*, Cs.—Buff-breasted Least Flycatcher.

Apparently a very rare species, as it was met with but on two occasions. At Inscription Rock, N. Mex., July 24, I observed a pair of old birds feeding the young. These latter were nearly full-fledged, and had evidently been raised in the immediate vicinity. In September a single immature bird was taken near Apache on a small brush-lined stream in a heavy pine-forest. Judging from the individuals seen, their habits differ in no noteworthy respect from those of the small flycatchers generally. The species was first described and introduced into our fauna by Dr. Coues, who gives it as a rare summer-resident at Fort Whipple, Ariz.

The plumage of the young differs from the adult in the paler fulvous of the under parts. There are two bands of strong fulvous across the wings; the tertiaries are edged externally with same, and also, with the secondaries, conspicuously tipped with ashy-white.

| No. | Sex. | Locality. | Date. | Collector. | Wing. | Tail. | Bill. | Tarsus. |
|---|---|---|---|---|---|---|---|---|
| 480 | ♂ ad. | Inscription Rock, N. Mex. | July 24 | Henshaw. | 2. 40 | 2. 16 | 0. 42 | 0. 54 |
| 481 | ♀ ad. | ......do ............ | July 24; | .... do .... | 2. 28 | 2. 09 | 0. 33 | 0. 53 |
| 482 | ♀ young | ......do ............ | July 24 | .... do .... | 2. 25 | 1. 71 | 0. 31 | 0. 54 |
| 483 | ♀ young | ......do ............ | July 24 | .... do .... | 2. 28 | 2. 09 | 0. 33 | 0. 53 |
| 679 | ♀ jun. | ......do ............ | July 24 | .... do .... | 2. 20 | 2. 00 | 0. 39 | 0. 61 |

114. *Pyrocephalus rubineus* (Bodd.), var. *mexicanus*, Scl.—Red Flycatcher.

This beautiful species was found to be not very uncommon in the valley of the Gila late in September. A specimen was secured here September 25 by Dr. Newberry, jr., who observed quite a number of others, which, owing to their shyness, could not be obtained. They were seen perching upon the mezquite-bushes, whence they darted constantly forth after insects.

| No. | Locality. | Date. | Collector. | Wing. | Tail. | Bill. | Tarsus. |
|---|---|---|---|---|---|---|---|
| A | Pueblo Viejo, Ariz........... | Sept. 25 | Newberry. | 3. 22 | 2. 58 | 0. 51 | 0. 62 |

ALCEDINIDÆ (the Kingfishers.)

115. *Ceryle alcyon,* (L.)—Belted Kingfisher.

An occasional individual seen on the small creeks and streams. Quite common on the Gila.

CAPRIMULGIDÆ (the Goatsuckers).

116. *Chordeiles popetue* (Vieill.), var. *henryi,* Cass.—Western Night-Hawk.

Abundant everywhere near streams and ponds.

| No. | Sex. | Locality. | Date. | Collector. | Wing. | Tail. | Bill. | Tarsus. |
|---|---|---|---|---|---|---|---|---|
| 774 | ♂ jun. | Gila River, Ariz ...... | Sept. 14 | Henshaw. | 7. 36 | 4. 18 | 0. 37 | 0. 63 |

117. *Chordeiles acutipennis* (Bodd.), var. *texensis*, Lawr.—Texas Night-Hawk.

Specimens obtained on the Gila River September 14, where it was abundant. Made its appearance perhaps half an hour before dusk, keeping over the river, where, in pursuit of insects, it flew swiftly in irregular circles. The common night-hawk was also present and associating freely with it, though the present species was the most abundant.

| No. | Sex. | Locality. | Date. | Collector. | Wing. | Tail. | Bill. | Tarsus. |
|-----|------|-----------|-------|-----------|-------|-------|-------|---------|
| 775 | ♂ jun. | Gila River, Ariz....... | Sept. 14 | Henshaw . | 6. 33 | 3. 86 | 0. 25 | 0. 50 |
| 776 | ♀ jun. | ...... do .............. | Sept. 14 | .... do .... | 6. 76 | 4. 22 | 0. 26 | 0. 50 |

118. *Antrostomus nuttallii*, (Aud.)—Nuttall's Poorwill.

In the whole extent of region traversed by the survey in Eastern Arizona, this whipporwill was found common. It was especially numerous near Apache and in the White Mountains, and I have heard them, in the latter part of August, half a dozen miles, singing soon after dusk within a short distance of each other. It begins to fly but a short time before dusk, and on this account is rarely met with and difficult to procure. Should a beaten road chance to pass through the forest, it will be found to be a favorite hunting-ground. I have often noticed that they make their first appearance in such a spot just before dusk, and remain in the neighborhood during the early evening. Probably the well-known abundance of flies and insects which frequent such places affords an explanation of this habit. The males continue their notes till very late in the season; for I frequently heard them during the first part of October, and even as late as the 17th. Young birds differ from the adult in having a lighter, purer shade of ash above and a suffusion of cinnamon over the back and wings. Below is a general fulvous tint, especially noticeable on the throat-patch.

| No. | Sex. | Locality. | Date. | Collector. | Wing. | Tail. | Bill | Tarsus. |
|-----|------|-----------|-------|-----------|-------|-------|------|---------|
| 612 | ♀ jun. | Apache, Ariz ......... | Aug. 25 | Henshaw . | 5. 63 | 3. 30 | 0. 40 | 0. 71 |
| 737 | ♀ ad. | Thirty miles south of Apache, Ariz. | Sept. 11 | .... do .... | 5. 80 | 3. 80 | 0. 42 | 0. 72 |
| 762 | ♀ ad. | ......do .............. | Sept. 12 | .... do .... | 5. 70 | 3. 47 | 0. 43 | 0. 69 |
| 763 | ♀ ad. | ......do .............. | Sept. 12 | .... do .... | 5. 87 | 3. 67 | 0. 43 | 0. 73 |
| 762a | ♀ ad. | ......do .............. | Sept. 12 | .... do .... | 5. 59 | 3. 48 | 0. 41 | 0. 73 |
| 874 | ♀ jun. | Camp Grant, Ariz .... | Sept. 26 | .... do .... | 5. 55 | 3. 66 | 0. 45 | 0. 68 |

CYPSELIDÆ (the Swifts).

119. *Panyptila saxatilis*, (Woodh.)—White-throated Swift.

This swift was found in considerable numbers flying over some lofty sandstone cliffs in the neighborhood of Wingate. They seem rarely to descend into the cañons and valleys, but pass to and fro above the highest points, keeping up a shrill twitter as they dart by with wonderful velocity. By ascending the highest points, we succeeded after some trouble in securing three specimens. In a large cave near by, I saw these birds enter crevices in the rocks with food for their young. In approaching the nests, they flew with scarcely abated speed, till just at the entrance, when their wings were closed, and they glided in with scarcely a perceptible stop, it almost appearing as though they literally flew in.

TROCHILIDÆ (the Hummingbirds).

120. *Stellula calliope*, Gld.—The Calliope-Hummingbird.

Though not nearly so abundant as either Rufous-backed or Broad-tailed Hummers, this diminutive species was still by no means rare. At Inscription Rock, N. Mex., where it was first seen, perhaps half a dozen were found in a two-days' stay. At Apache, during the latter part of August and 1st of September, it was rather common; but in the higher portions of the White Mountains, it was most abundant, and here, I doubt not, it finds its summer-home. At Camp Grant, the 27th of August, it was still present, though in small numbers.

| No. | Sex. | Locality. | Date. | Wing. | Tail. | Bill. |
|---|---|---|---|---|---|---|
| 470 | ♀ | Inscription Rock, N. Mex | July 23 | 1.60 | 1.97 | 0.60 |
| 474 | ♂ ad. | ......do | July 23 | 1.47 | 0.98 | 0.57 |
| 488 | ♀ ad. | ......do | July 24 | 1.64 | 0.95 | 0.62 |
| 490 | ♂ ad. | ......do | July 24 | 1.45 | 0.98 | 0.54 |
| 497 | ♂ ad. | ......do | July 24 | 1.48 | *0.98 | 0.60 |
| 537 | ♀ ad. | ......do | Aug. 9 | 1.63 | 0.94 | 0.63 |
| 538 | ♀ ad. | Apache, Ariz | Aug. 8 | 1.63 | 0.98 | 0.61 |
| 610 | ♂ jun. | ......do | Aug. 25 | 1.65 | 0.98 | 0.60 |
| 647 | ♂ ad. | ......do | Aug. 28 | 1.50 | 0.97 | 0.55 |
| 648 | ♂ jun. | ......do | Aug. 28 | 1.63 | 0.99 | 0.55 |
| 649 | ♂ jun. | ......do | Aug. 28 | 1.63 | 1.00 | 0.58 |
| 551 | ♀ ad. | White Mountains, Ariz | ......... | 1.60 | 0.95 | 0.58 |
| 552 | ♀ | ......do | ......... | 1.64 | ˙0.98 | 0.61 |
| 637 | ♂ ad. | ......do | ......... | 1.47 | 0.91 | 0.55 |
| 849 | ♂ jun. | Camp Grant, Ariz | ......... | 1.59 | 0.91 | 0.57 |

121. *Trochilus alexandri*, Bourc. & Muls.—Black-chinned Hummingbird.

Two specimens of this hummer were taken at Apache in August, and a third at Camp Grant in September. It is thus apparently rare in this portion of Arizona, in which Territory it has not hitherto been detected.

| No. | Sex. | Locality. | Date. | Collector. | Wing. | Tail. | Bill. |
|---|---|---|---|---|---|---|---|
| 521 | ♂ jun. | Apache, Ariz | Aug. 5 | Henshaw | 1.77 | 1.23 | 0.80 |
| 587 | ♂ jun. | ......do | Aug. 21 | ....do .... | 1.68 | 1.05 | 0.72 |
| 844 | ♂ jun. | Camp Grant, Ariz | Sept. 24 | ....do .... | 1.90 | 1.17 | 0.77 |

122. *Calypte annæ*, (Lesson.)—Anna Hummingbird.

As this species has been found within our borders only in the coast-region of California, its detection at Camp Grant, Ariz., has widely extended its distribution. It is likely that it inhabits the intermediate region in greater or less numbers. At the point where it was found it is by no means rare, as I saw in the neighborhood of twenty during the four days spent in collecting in this vicinity. They were always seen in the immediate vicinity of the creeks, where only at this late season there remained a few of the bright flowers about which they were seen hovering. Their large size rendered them very conspicuous among the other species, and as if aware of this they were much the shyest of all.

| No. | Sex. | Locality. | Date. | Collector. | Wing. | Tail. | Bill. |
|-----|------|-----------|-------|------------|-------|-------|-------|
| 852 | ♂ | Camp Grant, Ariz.......... | Sept. 24 | Henshaw. | 2.00 | 1.16 | 0.68 |
| 853 | ♂ ad. | ......do..................... | Sept. 24 | ....do .... | 1.94 | 1.15 | 0.73 |
| 854 | ........ | ......do..................... | Sept. 24 | ....do .... | 1.90 | 1.37 | 0.71 |
| 873 | ♂ | ......do..................... | Sept. 27 | ....do .... | 1.93 | 1.25 | 0.68 |
| 879 | ♂ jun. | ......do..................... | Sept. 27 | ....do .... | 1.95 | 1.30 | 0.70 |

123. *Selasphorus rufus*, (Gmel.)—Rufous-backed Hummingbird.

By far the most abundant of the family in New Mexico and Arizona, as shown in every locality visited by our party. Quite numerous at Inscription Rock, but at Apache during the month of August they were seen literally by hundreds hovering over the beds of brightly-tinted flowers, which in the mountains especially grow in the greatest profusion on the borders of the mountain-streams. This bird seems to affect no particular locality, but is about equally abundant on the high mountains, in the open tracts of pine-woods, in the valleys and deep cañons, or, in fact, wherever flowers are found. The males are very pugnacious, and wage unremitting warfare on all the other species, as well among themselves. Even as late as August it was not uncommon to see these birds still in pairs, and established in certain areas, of which they appeared to consider themselves the sole possessors, allowing no intruders. They manifested an especial animosity against the Broad-tailed Hummer, and, on the appearance of one, would instantly dart forth with shrill, angry notes, and attack and drive away the intruder, while the female, sitting on some neighboring tree, would watch the oft-repeated contest with evident interest and solicitude. At Camp Grant, during the last days of September, they were still numerous, but after leaving this point I did not again see the species. A series of over forty specimens were secured, representing all stages of plumage.

| No. | Locality. | Date. | Collector. | Wing. | Tail. | Bill. |
|-----|-----------|-------|------------|-------|-------|-------|
| 38 | Deer Spring, N. Mex .............. | July 25 | Newberry. | 1.63 | 1.33 | 0.65 |
| 491 | Inscription Rock, N. Mex............ | July 24 | Henshaw . | 1.80 | 1.31 | 0.68 |
| 495 | ...... do ..................... | July 24 | ....do .... | 1.75 | 1.25 | 0.72 |
| 496 | ......do . ..................... | July 24 | ....do .... | 1.94 | 1.50 | 0.72 |
| 536 | Mountains near Apache, Ariz....... | July 24 | ....do .... | 1.60 | 1.30 | 0.34 |
| 537 | ......do ..................... | Aug. 5 | ....do .... | 1.63 | 1.30 | 0.68 |
| 559 | ......do ..................... ... | Aug. 9 | ....do .... | 1.60 | 1.32 | 0.65 |
| 563 | ......do ..................... | Aug. 9 | ....do .... | 1.59 | 1.27 | 0.62 |
| 564 | ......do ..................... | Aug. 9 | ....do .... | 1.58 | 1.23 | 0.66 |
| 603 | Apache, Ariz ..................... | Aug. 11 | ....do .... | 1.73 | 1.15 | 0.64 |
| 616 | ......do ..................... | Aug. 23 | ....do .... | 1.74 | 1.25 | 0.71 |
| 624 | ......do ..................... | Aug. 27 | ....do .... | 1.80 | 1.28 | 0.66 |
| 888 | Camp Grant, Ariz................. | Sept. 27 | ....do .... | 1.70 | 1.26 | 0.73 |

124. *Selasphorus platycercus*, (Sw.)—Broad-tailed Hummingbird.

Found at same time and in same localities as the preceding, but not nearly so abundant. It is, however, very numerous, and in certain localities, as at Inscription Rock, exceeded the Rufous-backed in numbers. Like it, seems to frequent no especial locality, but follows the range of the flowers everywhere.

| No. | Sex. | Locality. | Date. | Collector. | Wing. | Tail. | Bill. |
|-----|------|-----------|-------|------------|-------|-------|-------|
| 91  | ♂ ad. | Apache, Ariz ............... | Sept. — | Newberry. | 1.88 | 1.36 | 0.75 |
| 492 | ♀ ad. | Inscription Rock, N. Mex .. | July 24 | Henshaw . | 2.02 | 1.38 | 0.76 |
| 539 | ♂ ad. | White Mountains, Ariz .... | Aug. 9 | ....do .... | 1.90 | 1.48 | 0.73 |
| 540 | ♂ ad. | Apache, Ariz ............... | Aug. 9 | ....do .... | 1.91 | 1.38 | 0.71 |
| 543 | ♂ ad. | White Mountains, Ariz...... | Aug. 8 | ....do .... | 1.95 | 1.33 | 0.71 |
| 568 | ♂ ad. | ......do .................... | Aug. 12 | ....do .... | 1.95 | 1.38 | 0.64 |
| 649 | ♂ ad. | ......do .................... | Aug. 28 | ....do .... | 1.90 | 1.33 | 0.64 |
| 873 | ♂ ad. | Camp Grant, Ariz ......... | Sept. 26 | ....do .... | 1.98 | 1.42 | 0.67 |

125. *Eugenes fulgens*, (Sw.)

SP. CHAR.—Male: tail rather deeply emarginated; head above violet purple; rest of upper parts bronzed green, becoming pure bronze on the tail; gorget brilliant emerald-green, with strong purple reflections; lower portion of breast and abdomen opaque black, more velvety toward the green of throat; sides of body dull green; wing above and below dull purple; upper and lower wing-coverts green; crissum pale brownish-gray; bill and feet black. Female: tail double-rounded, above dark metallic-green, each feather edged with ash, below dull white; feathers of throat and fore part of breast with dull grayish-green centers; sides green, edged with ash; wing dull purple; each feather of the tail except the two central, which are green throughout, with broad purple band; three outer tail-feathers broadly tipped with dull white, which, on the outer, extends slightly further up on the outer web. Length, 4.61; wing, 2.43; tail, 1.75; bill, 1.09.

This fine species is now for the first time introduced into our fauna, it never having been before observed farther north than the table-lands of Central Mexico. A female was taken on a small stream issuing from the mountains, at Camp Grant, Ariz., September 24, and identified by Mr. G. N. Lawrence as of this species. When first seen, it was being pursued by another hummer, of which I obtained but a glimpse as they darted past through the trees, but I have little doubt that it was a second of the same species. I think it not unlikely that the species will be found to be not uncommon in the mountainous districts of Southern Arizona and New Mexico.

CUCULIDÆ (the Cuckoos).

126. *Geococcyx californianus*, (Lesson.)—Road-Runner; Chaparral-Cock.

From information obtained from hunters and guides, this remarkable bird seems to be rather common and generally distributed through the valley of the Gila and southward. It is of retiring habits, and, as it keeps in the chaparral and thickets, is very likely to be overlooked. I secured one of these birds on the Gila in October. It was sitting quietly on a stump, sunning itself in the rays of the rising sun, and had allowed several pack-mules to pass within twenty feet without manifesting alarm. The moment, however, it found itself observed, it dropped down, and ran swiftly a short distance, till an uprising bank hid it from view, when it stopped, and I overtook and shot it. The crop was nearly filled with grasshoppers and a few coleopterous insects. Dr. Newberry also procured a specimen at Fort Bowie.

PICIDÆ (the Woodpeckers).

127. *Picus villosus*, L., var. *harrisii*, Aud.—Western Hairy Woodpecker.

The most abundant of its tribe in the region visited by the survey. With a decided preference for the pines, it yet, in the fall, is found straggling all over the country and frequenting the deciduous trees generally.

128. *Picus pubescens* L., var. *gairdneri*, Aud.—Western Downy Woodpecker.

One or two noticed among the cottonwoods along the Gila River in October. The rarity of this species, as compared with the extreme abundance of the preceding, is very remarkable.

129. *Picus scalaris*, Wagl.—Ladder-backed Woodpecker.

Not found at Apache. In a cañon thirty miles south, Dr. Newberry shot one and saw several others. Along the Gila and San Pedro Rivers it appeared to be a rather common woodpecker, noticed most often about the mezquite, the trunks of which it appeared to scan most carefully for food. Its notes and manners are much like those of the Downy Woodpecker.

| No. | Sex. | Locality. | Date. | Collector. | Wing. | Tail. | Bill. | Tarsus. |
|-----|------|-----------|-------|-----------|-------|-------|-------|---------|
| 736 | ♀ jun. | Thirty miles south of Apache, Ariz. | Sept. 11 | Newberry. | 4.00 | 2.71 | 0.80 | 0.70 |
| 900 | ♀ | San Pedro, Ariz....... | Oct. 3 | Henshaw . | 3.85 | 2.62 | 0.77 | 0.69 |
| 901 | ♀ | ......do ............. | Oct. 3 | Magnet... | 3.88 | 2.63 | 0.73 | 0.67 |
| 912 | ♀ | Gila River, Ariz ...... | Oct. 15 | Henshaw . | 3.86 | 2.69 | 0.81 | 0.68 |

130. *Sphyropicus varius* (L.), var. *nuchalis*, Bd.—Red-naped Woodpecker.

An abundant and generally well-distributed species, found among the deciduous trees.

131. *Sphyropicus thyroideus*, Cass.—Brown-headed Woodpecker; Black-breasted Woodpecker; Williamson's Woodpecker.

Specimens secured near the headwaters of the Gila, in New Mexico. As in summer, found only in the pine-woods, where they associated with the bands of Nuthatches.

| No. | Sex. | Locality. | Date. | Collector. | Wing. | Tail. | Bill. | Tarsus. |
|-----|------|-----------|-------|-----------|-------|-------|-------|---------|
| 977 | ♂ ad. | Gila River, N. Mex.... | Nov. — | Henshaw. | 5.29 | 3.73 | 0.93 | 0.86 |
| 980 | ♂ ad. | ......do ............. | Nov. 5 | ....do .... | 5.43 | 3.70 | 1.00 | 0.85 |
| 981 | ♂ ad. | ......do ............. | Nov. 5 | ....do .... | 5.47 | 3.93 | 0.90 | 0.82 |

132. *Centurus uropygialis*, Bd.—Gila Woodpecker.

Not met with farther north than the valley of the Gila. Here, however, and to the southward, it was not uncommon. The Giant Cactus (*Cereus giganteus*), which forms a most striking and characteristic feature in this region, bears all over its body marks of the work of these birds, large patches being dug entirely out, as though the pith or sap was sought for. Its trunk, too, appears to afford a favorite nesting-site, and the excavations for this purpose are often to be seen. On the San Pedro, this species was found frequenting the mezquite-trees. They were everywhere very shy.

| No. | Sex. | Locality. | Date. | Collector. | Wing. | Tail. | Bill. | Tarsus. |
|---|---|---|---|---|---|---|---|---|
| 890 | ♀ ad. | San Pedro, Ariz. ..... | Oct. 2 | Henshaw . | 5. 30 | 3. 83 | 1. 05 | 0. 91 |
| 103 | ♂ ad. | Pueblo Viejo, N. Mex . | Sept. 27 | Newberry. | 5. 40 | 3. 87 | 1. 28 | 0. 88 |
| 44 | ♂ ad. | ......do ............... | Sept. 19 | ....do .... | 5. 10 | 3. 52 | 1. 17 | 0. 95 |

133. *Melanerpes torquatus*, (Wils.)—Lewis's Woodpecker.

A few only seen about Apache, where it was noticed circling about the tops of the high, isolated pines. Dr. Newberry, jr., reported it as quite numerous in the White Mountains, where it kept constantly among the pines.

| No. | Sex. | Locality. | Date. | Collector. | Wing. | Tail. | Bill. | Tarsus. |
|---|---|---|---|---|---|---|---|---|
| 700 | ♀ jun. | Apache, Ariz.......... | Sept. 5 | Henshaw. | 6. 55 | 4. 41 | 1. 05 | 0. 93 |

134. *Melanerpes formicivorus*, (Sw.)—Californian Woodpecker.

In the neighborhood of Apache, in August, this was much the most abundant of the woodpeckers. They showed extreme sociability, moving about in small companies of from six to twelve, and keeping constantly in the oak-groves. When not busied gleaning insects among the branches, they were constantly engaged in playing, chasing each other in and out among the trees, and apparently the utmost good-will prevailed in their companies.

| No. | Sex. | Locality. | Date. | Collector. | Wing. | Tail. | Bill. | Tarsus. |
|---|---|---|---|---|---|---|---|---|
| 86 | ♀ | Oak Orchard, Ariz .... | Sept. 1 | Newberry. | 5. 17 | 3. 12 | 1. 10 | 0. 75 |
| 529 | ♀ ad. | Apache, Ariz.......... | Aug. 6 | Henshaw . | 5. 45 | 3. 45 | 0. 93 | 0. 77 |
| 592 | ♂ jun. | ......do .............. | Aug. 22 | ....do .... | 5. 60 | 3. 63 | 0. 98 | 0. 77 |
| 593 | ♂ jun. | ......do .............. | Aug. 22 | ....do .... | 5. 47 | 3. 54 | 0. 93 | 0. 79 |
| 641 | ♀ ad. | ......do .............. | Aug. 28 | ....do .... | 5. 65 | 3. 70 | 0. 90 | 0. 76 |
| 672 | ♂ | ......do .............. | Sept. 1 | ....do .... | 5. 60 | 3. 45 | 1. 02 | 0. 79 |
| 686 | ♂ | ......do .............. | Sept. 2 | ....do .... | 5. 45 | 3. 40 | 1. 10 | 0. 81 |
| 731 | ♀ jun. | ......do ...... ...... | Sept. 10 | ....do .... | 5. 54 | 3. 30 | 1. 02 | 0. 80 |

135. *Colaptes auratus* (L.), var. *mexicanus*, Sw.—Red-shafted Woodpecker.
Numerous, without reference to special locality.

| No. | Sex. | Locality. | Date. | Collector. | Wing. | Tail. | Bill. | Tarsus. |
|---|---|---|---|---|---|---|---|---|
| 911 | ♂ ad. | San Francisco River, Ariz. | Oct. 14 | Henshaw. | 6. 37 | 4. 76 | 1. 63 | 1. 13 |
| 913 | ♀ | ......do .............. | Oct. 16 | ....do .... | 6. 45 | 4. 90 | 1. 40 | 1. 07 |

STRIGIDÆ (the Owls).

**136.** *Otus brachyotus,* (Gmel.)—Short-eared Owl.

A single specimen obtained near Camp Bowie, Southeastern Arizona, which was the only occasion the species was met with. This bird was started from a low clump of bushes on an open plain, and flew in a wild, uncertain manner, as though completely bewildered. It proved, however, no easy matter to get within gun-shot of it a second time, and several unsuccessful attempts were made ere a long shot brought it down.

**137.** *Scops asio* (L.), var. *maccalli,* Cass.—Western Mottled Owl.

This bird was very common, both in Arizona and New Mexico, and is, I think, the most numerous of the family in this region. Whenever our camp chanced to be made near one of the groves of oaks, which are numerous, these owls were sure to be heard soon after dusk, and, not infrequently, several would take up their stations in a tree within a few feet of the camp-fire, and remain for an hour or more, apparently to satisfy their curiosity, uttering, from time to time, their low, responsive cries. Their notes vary much in length, but, when full, consist of two prolonged syllables, with quite an interval between, followed by a rapid utterance of six or seven notes, which, at the end, are run together. They are very sociable in their disposition, and as soon as fairly dusk the first call of a solitary bird may be heard issuing from some thicket, where it has remained in concealment during the day. After one or two repetitions, this will be answered by another, perhaps half a mile away, and soon by a third and a fourth, and, as apparent, all coming together; and I have heard at least eight of these owls, all congregated within a short distance in the tree-tops. When the band was complete, they would move off, still apparently keeping together, till their notes were lost in the distance.

| No. | Sex. | Locality. | Date. | Collector. | Wing. | Tail. | Bill. | Tarsus. |
|---|---|---|---|---|---|---|---|---|
| 790 | ♂ ad. | Gila River, Ariz....... | Sept. 15 | Henshaw. | 6.58 | 3.45 | 0.53 | 1.32 |
| 895 | ♂ | Camp Grant, Ariz..... | Sept. 29 | ....do .... | 6.62 | 3.45 | 0.60 | 1.15 |
| 907 | ♀ | San Pedro, Ariz ....... | Oct. 4 | ....do .... | 6.48 | 3.58 | 0.54 | 1.25 |

**138.** *Scops flammeola,* Licht.—Flammulated Owl.

This rare species has hitherto been known to our fauna through a single specimen taken at Fort Crook, Cal. I think, however, that it may be not uncommon in Arizona, though, like others of this genus, its strictly nocturnal habits render it extremely liable to escape detection. A fine specimen was secured by Dr. Newberry, jr., in a cañon thirty miles south of Apache. Having shot a small bird, he was pushing through the brush to pick it up, when this little owl started from a low tree, where it was concealed, probably asleep, and, alighting a few yards distant, he shot it. At the report of his gun, a second flew out from a low bush, but was lost in the thick brush. The following evening, when returning to camp, gun in hand, I was imitating the notes of the Screech-Owl, and was answered by notes similar in character, but shorter and weaker. Stationing myself directly under an oak, the top branches of which I could see outlined against the sky, and continuing the call, I soon saw the form of a diminutive owl clearly defined against the sky, and I think it probably was the mate of the one shot. Upon shooting, the bird fell part way down, but, recovering itself, I obtained a second's glimpse of it as it flew out, and was lost in the deep shadows of the cañon's sides.

| No. | Locality. | Date. | Collector. | Wing. | Tail. | Bill. | Tarsus. |
|---|---|---|---|---|---|---|---|
| 735 | Thirty miles south of Apache, Ariz. | Sept. 11 | Newberry. | 5.28 | 2.73 | 0.88 | 0.82 |

139. *Bubo virginianus* (Gm.), var. *arcticus*, Swains.—Western Great Horned Owl.

The only specimen taken is quite typical of the paler, grayer race, which represents the Horned Owl in the West. The species was very abundant, and scarcely a camp was made but we were made aware of the presence of these owls by their loud hootings through the night. During the day they remain hidden in the deep, dark cañons, or among the thick foliage of the largest cottonwoods.

| No. | Sex. | Locality. | Date. | Collector. | Wing. | Tail. | Bill. | Tarsus. |
|---|---|---|---|---|---|---|---|---|
| 793 | ♀ ad. | Gila River, Ariz. | Sept. 17 | Henshaw. | 14.25 | 8.50 | 1.66 | 0.20 |

140. *Glaucidium passerinum* (L.), var. *californicum*, Scl.—Californian Pigmy Owl.

This little owl is apparently quite common in Arizona and New Mexico. It does not appear to be at all a nocturnal species, but was observed to be most active in the early morning and late afternoon, and on one occasion was seen flying at broad noonday. Their notes are quite similar to those of the Mottled Owl (*scopsmaccalli*), by imitating which I succeeded in enticing one, step by step, till he finally sat on the top of a small oak within thirty feet, and scanned my person with evident astonishment, and, I could not help fancying, with an air of abused confidence.

A young bird collected by Dr. Newberry, jr., is quite appreciably different from the adult. The entire plumage has more of a slaty tinge, while the back and under parts are strongly suffused with rufous. The head above lacks the numerous rounded reddish-white spots, but each feather has a single elongated white spot at the tend of the shaft.

| No. | Sex. | Locality. | Date. | Collector. | Wing. | Tail. | Bill. | Tarsus. |
|---|---|---|---|---|---|---|---|---|
| 46 | ♀ jun. | Near Apache, Ariz .... | Aug. 9 | Newberry. | 3.82 | 2.83 | 0.45 | 0.78 |
| 761 | ♂ | Thirty miles south of Apache, Ariz. | Sept. 12 | Henshaw . | 3.58 | 3.05 | 0.40 | 0.76 |
| 971 | ♂ ad. | Gila River, Ariz. | Oct. 26 | .... do .... | 3.87 | 3.05 | 0.47 | 0.88 |

141. *Speotyto cunicularia* (Mol.), var. *hypugæa*, Bonap.—Burrowing Owl.

This curious owl appears not to be a very abundant resident either in Arizona or New Mexico, at least in those portions visited by the survey during the past season. They prefer the lower plains, and are not found, I think, at a higher altitude than 6,000 feet. Near Zuñi, N. Mex., Forts

Grant and Bowie, Ariz., and a few other places, they were seen always about the settlements of the prairie dogs. Their sight in the day appears to be remarkably good, and, as all I saw were very shy, it proved to be no easy matter to get within shooting-distance. Their flight is rather laborious and irregular, and they do not fly to any great distance when alarmed, but try to hide in the mouths of the prairie-dog-holes, though I never saw one take refuge in them.

| No. | Sex. | Locality. | Date. | Collector. | Wing. | Tail. | Bill. | Tarsus. |
|-----|------|-----------|-------|-----------|-------|-------|-------|---------|
| 933 | ♀ | Camp Bowie, Ariz..... | Oct. 9 | Henshaw . | 6.66 | 3.40 | 0.55 | 1.73 |

FALCONIDÆ (the Falcons).

142. *Falco saker* (Schleg.), var. *polyagrus*, Cass.—Prairie-Falcon.

Seen at several points in Arizona and New Mexico. In habits, shy and solitary. I never observed it hunting its prey, but when flying it maintains a direct course through the air from point to point, and progresses very swiftly by short, powerful strokes of the wings. Its flight is sufficiently peculiar to distinguish it from any other hawk with which I am acquainted.

Bill bluish, black at tip ; feet lead-color.

| No. | Sex. | Locality. | Date. | Collector. | Wing. | Tail. | Bill. | Tarsus. |
|-----|------|-----------|-------|-----------|-------|-------|-------|---------|
| 974 | ♂ jun. | Gila River, N. Mex .... | Nov. 1 | Henshaw . | 11.55 | 7.00 | 0.77 | 1.85 |

143. *Falco columbarius*, L.—Pigeon-Hawk.

I am quite confident that I saw this species once or twice in Southern Arizona, and at least once in New Mexico, on the upper sources of the Gila. It is given by Dr. Coues, in his list of "Birds of Fort Whipple," as a common resident.

144. *Falco femoralis*, Temm.—Aplomado-Falcon.

A hawk was seen in a mountainous locality near Camp Bowie, Southeastern Arizona, which was without doubt of this species. It was about the size of a Cooper's Hawk, and, as it passed rapidly by within fair shooting-distance, the black band across the abdomen was very conspicuous. It has twice been taken on our southern border.

145. *Falco sparverius*, L.—Sparrow-Hawk.

Common everywhere, both in Arizona and New Mexico, breeding in the holes of trees along the streams. Seems rarely to trouble the small birds, but lives almost exclusively upon grasshoppers, of which it finds a great abundance.

| No. | Sex. | Locality. | Date. | Collector. | Wing. | Tail. | Bill. | Tarsus. |
|-----|------|-----------|-------|-----------|-------|-------|-------|---------|
| 586 | ♂ ad. | Camp Apache, Ariz.... | Aug. 4 | Henshaw . | 7.22 | 5.23 | 0.45 | 1.45 |
| 73 | ♀ jun. | ......do ............. | Sept. 13 | Magnet... | 7.44 | 5.55 | 0.50 | 1.40 |

146. *Pandion haliaëtus* (L.), var. *carolinensis*, Gmel.—Fish-Hawk.

An occasional Fish-Hawk was seen busily employed in its vocation on the small streams. On the Gila, however, which is plentifully stocked with fish, it seems to find a congenial home, and is quite abundant along its banks.

| No. | Sex. | Locality. | Date. | Collector. | Wing. | Tail. | Bill. | Tarsus. |
|---|---|---|---|---|---|---|---|---|
| 938 | ♂ | Gila River, N. Mex.... | Sept. 17 | Magnet .. | 18.00 | 8.50 | 1.13 | 2.08 |

147. *Circus cyaneus* (L.), var. *hudsonius*, L.—Marsh-Hawk.

A common hawk in Arizona and New Mexico, but here, as elsewhere, confined rather exclusively to the vicinity of marshes and water-courses. I noticed many on the creeks about Camp Grant, where they were remarkably tame and unsuspicious.

| No. | Sex. | Locality. | Date. | Collector. | Wing. | Tail. | Bill. | Tarsus. |
|---|---|---|---|---|---|---|---|---|
| 851 | ♀ jun. | Camp Grant, Ariz..... | Sept. 24 | Henshaw. | 13.40 | 8.60 | 0.63 | 2.86 |

148. *Nisus fuscus*, (Gmel.)—Sharp-shinned Hawk.

Quite common, both in Arizona and New Mexico. Preys much upon small birds; the doves (*Zenaidura carolinensis*) suffering much from its attacks.

| No. | Sex. | Locality. | Date. | Collector. | Wing. | Tail. | Bill. | Tarsus. |
|---|---|---|---|---|---|---|---|---|
| 867 | ♂ jun. | Camp Grant, Ariz .... | Sept. 24 | Henshaw. | 6.88 | 5.72 | 0.40 | 1.90 |
| 875 | ♂ jun. | ......do............... | Sept. 27 | .... do .... | 6.80 | 5.90 | 0.40 | 1.80 |
| 914 | ♀ ad. | Gila River, Ariz ...... | Oct. 16 | .... do .... | 8.10 | 6.60 | 0.53 | 2.17 |

149. *Nisus cooperi*, (Bon.)—Cooper's Hawk.

An abundant species throughout Eastern Arizona and Western New Mexico. While sitting in my tent one day at Apache, I noticed one of these hawks making repeated attacks upon a raven. It would force the raven to take refuge in a tree, and then fly to some neighboring perch and take its stand. The moment the persecuted raven essayed to move away, the hawk flew out and swooping down upon it struck it and again forced it to cover. This was repeated several times, and apparently for no other reason than for the amusement of the hawk, though, judging from the discontented squawks and cries which the abused raven gave vent to, the pleasure was by no means mutual. So engrossed was the falcon in this sport that it allowed me unnoticed to walk up within a few feet, when my gun settled the dispute.

Bill bluish-black; legs and feet yellow.

| No. | Sex. | Locality. | Date. | Collector. | Wing. | Tail. | Bill. | Tarsus. |
|---|---|---|---|---|---|---|---|---|
| 616 | ♀ jun. | Apache, Ariz........... | Aug. 26 | Henshaw . | 10.30 | 9.75 | 0.68 | 2.65 |
| 802 | ♀ jun. | Goodwin, Ariz ........ | Sept. 20 | .... do .... | 10.50 | 9.75 | 0.70 | 2.64 |
| 899 | ♀ jun. | San Pedro, Ariz. ...... | Oct. 1 | .... do .... | 10.70 | 9.75 | 0.71 | 2.58 |

**150.** *Buteo swainsoni*, Bon.—Swainson's Hawk.

At Camp Grant, Ariz., in the latter part of September, this hawk was present in very large numbers, and they seemed to have centered in this spot from the surrounding country. About a mile below the post, out on the plain, the stream was bordered by some large cottonwoods, and these were habitually used as roosting-places by the turkey-buzzards and hawks conjointly, as the whitened appearance of the branches and the ground below testified, as well as the fetid odor in their vicinity. Hawks and buzzards appeared to be on terms of the most intimate companionship with each other, and one tree often held seven or eight of either birds. The buzzards seemed if anything rather the shyer of the two, and were generally the first to start, when immediately the whole band would leave their perches, and begin circling in the air, gradually ascending higher and higher till out of danger, and thus continue wheeling about till the coast was clear, when all would again resume their perches. After leaving these, and getting fairly on the wing, which they did rather clumsily, the flight of this hawk is firm and easy, and as they gradually soar higher and higher in circles, their flight bears no little resemblance to that of the buzzards, though it is less powerful and not so well sustained. Indeed, when thus mingled with the buzzards, the general resemblance is rather striking. I am not aware that these hawks feed upon carrion, though that they occasionally do so is not unlikely. The crops of all those shot were found fairly crammed with grasshoppers; and as these insects were very abundant, the hawks, as a matter of course, were very fat.

| No. | Sex. | Locality. | Date. | Collector. | Length. | Stretch. | W. | T. | C. | Tar. |
|-----|------|-----------|-------|-----------|---------|----------|-----|-----|-----|------|
| 870 | ♀ ad. | Camp Grant, Ariz. | Sept. 26 | Henshaw. | 18.86 | 47.74 | 15.31 | 9.32 | 0.84 | 2.12 |

Bill black ; cere yellow ; legs and feet yellow.

| No. | Sex. | Locality. | Date. | Collector. | Wing. | Tail. | Cul. | Tarsus. |
|-----|------|-----------|-------|-----------|-------|-------|------|---------|
| 871 | ♀ ad. | Camp Grant, Ariz ..... | Sept. 26 | Henshaw. | 15.15 | 8.50 | 0.84 | 2.70 |

| No. | Sex. | Locality. | Date. | Collector. | Length. | Stretch. | W. | T. | C. | Tar. |
|-----|------|-----------|-------|-----------|---------|----------|-----|-----|-----|------|
| 876 | ♂ jun. | Camp Grant, Ariz. | Sept. 26 | Henshaw. | 18.74 | 47.24 | 14.74 | 8.62 | 0.80 | 2.49 |

Iris brown; cere greenish-yellow; base lower mandible and edge along gape greenish-yellow; legs and feet yellow.

| No. | Sex. | Locality. | Collector. | Length. | Stretch. | W. | T. | C. | Tar. |
|-----|------|-----------|-----------|---------|----------|-----|-----|-----|------|
| 877 | ♂ ad. | Camp Grant, Ariz ...... | Henshaw. | 19.00 | 47.86 | 15.00 | 8.49 | 0.85 | 2.49 |

Iris dark-brown ; cere greenish-yellow; legs and feet yellow.

| No. | Sex. | Locality. | Date. | Collector. | Wing. | Tail. | Bill. | Tarsus. |
|---|---|---|---|---|---|---|---|---|
| 894 | ♀ ad. | Camp Grant, Ariz..... | Sept. 29 | Henshaw. | 16.20 | 8.50 | 1.00 | 2.78 |

151. *Buteo borealis* (Gmel.), var. *calurus*, Cassin.—Western Red-tailed Hawk.

A wide-spread species, seen occasionally during the season.

| No. | Sex. | Locality. | Date. | Collector. | Wing. | Tail. | Bill. | Tarsus. |
|---|---|---|---|---|---|---|---|---|
| 687 | ♂ ad. | Apache, Ariz.......... | Sept. 1 | Henshaw. | 14.65 | 9.50 | 1.00 | 3.27 |

152. *Haliaëtus lucocephalus* (L.)—American Eagle; Bald Eagle.

An adult pair of these magnificent birds was seen in a cañon a few miles south of Apache. A solitary bird was to be seen now and then, perched on some lofty dead stub overlooking a stream, and apparently on the watch for fish. As Fish-Hawks are by no means numerous, the eagles are compelled to have recourse to their own resources, and do their own fishing and hunting to a much greater extent than upon the coast, where their successful attempts at robbing the Osprey are well known. Among the Zuni Indians, these birds are highly prized, as affording the feathers with which they deck themselves at their sacred feasts and dances. At the pueblo Zuni, I saw perhaps a dozen kept in wicker-inclosures. They presented a most lamentable appearance, as their bodies were devoid of feathers, which had been plucked out long before. The quills and tail-feathers are especially valued.

CATHARTIDÆ (the American Vultures).

153. *Rhinogryphus aura* (L.)—Red-headed Vulture.

The turkey-buzzard is found throughout Eastern Arizona and Western New Mexico, where it congregates on the outskirts of the settlements, and feeds upon the refuse and carrion which may fall in its way. At Apache and Camp Grant it was particularly numerous, and at the latter place freely associated with the Swainson's Hawk (*B. swainsonii*). The quills of this bird are generally used by the Indians to feather their arrows.

COLUMBIDÆ (the Pigeons).

153. *Columba fasciata*, Say.—Band-tailed Pigeon.

Of the habits of this beautiful pigeon I am able to add nothing to what is already known. At Apache, Ariz., further north than which I did not meet with it, I obtained a single specimen, August 21. In passing southward from here, during September I saw an occasional flock, on one occasion at least two hundred. They were generally engaged in picking up seeds, and betrayed the utmost shyness, so that I found it impossible to either observe their habits or procure specimens.

Iris red ; bill yellow, black at tip ; legs and feet yellow.

| No. | Sex. | Locality. | Date. | Collector. | Wing. | Tail. | Bill. | Tarsus. |
|-----|------|-----------|-------|------------|-------|-------|-------|---------|
| 582 | ♂ ad. | Apache, Ariz.......... | Aug. 21 | Henshaw. | 8.80 | 6.60 | 0.62 | 0.98 |

154. *Melopelia leucoptera,* (L.)—White-winged Dove.

A single bird obtained on the Gila River, in New Mexico, and the only one seen. I am, therefore, inclined, to consider it as a rare species, though probably occurring in Eastern Arizona.

| No. | Sex. | Locality. | Date. | Collector. | Wing. | Tail. | Bill. | Tarsus. |
|-----|------|-----------|-------|------------|-------|-------|-------|---------|
| 973 | ♂ | Gila River, N. Mex.... | Oct. 28 | Henshaw. | 6.23 | 5.32 | 0.79 | 0.90 |

155. *Zenaidura carolinensis,* (L.)—Common Dove.

Abundant throughout Eastern Arizona. Being rarely molested, they seem to have no fear of man, and at Apache, where they were especially numerous, were accustomed to remain about our camp all day.

MELEAGRIDÆ (the Turkeys).

156. *Meleagris gallopavo* (L.)—Mexican Turkey.

The wild turkey is found abundantly from Apache throughout the mountainous portion of Southeastern Arizona. In New Mexico it was met with further to the north, in the mountains, and I was informed by Colonel Alexander that he had found them in large numbers in the Raton Mountains, in extreme Northern New Mexico. It breeds abundantly through the White Mountains, Arizona, and about the middle of August several broods of the young, about two-thirds grown, were met with. Toward the head of the Gila, in New Mexico, the cañons, in November, were found literally swarming with these magnificent birds; in many places the ground being completely tracked up where they had been running. As many as eleven were killed by the members of a party during a day's march. They roost at night in the large cottonwoods by the streams, and soon after daylight, having visited the stream, they usually betake themselves to the dry hills, where they feed, in the fall, at least, almost exclusively upon the seeds of grasses and grasshoppers. I think they return once or twice during the day to drink, the dry nature of their food rendering a copious supply of water necessary. In these wilds, they appear to be wholly unsuspicious, and without knowledge of danger from man, and, if not shot at, will allow one to get within a few yards without manifesting any distrust. They rarely fly, except when very hard pressed, but, when alarmed, run with such rapidity as to quickly outstrip the fleetest foot, betaking themselves to the steep sides of the ravines, which they easily scale, and soon elude pursuit. Apparently, the only dangers they have to fear in these regions are from birds of prey, which attack the young, but more especially from the panthers. In certain portions of the Gila Cañon, the tracks of these animals were very numerous, and always these sections appeared to have been entirely depopulated of Turkeys, an occasional pile of feathers marking the spot where one had fallen a victim to one of these animals. The molt is protracted till very late, as, though completed by November 15, many of the feathers were but partially developed, with the stems still soft.

TETRAONIDÆ (the Grouse).

157. *Canace obscura* (Say).—Dusky Grouse.

A rather common inhabitant of the White Mountains, Arizona. Quite a number were shot in August, and on the 15th Dr. Newberry, jr., saw a female with young, probably a second brood. This locality is much further to the south than the bird was known to range. I think, however, that it will be likely to be found extending in the mountains well down to our southern border.

PERDICIDÆ (the Partridges).

158. *Lophortyx gambelii*, Nutt.—Gambel's Partridge.

Met with by Dr. Newberry, jr., a few miles south of Santa Fé. It here, however, is not nearly so abundant as to the southward, in Arizona and New Mexico. At Apache, they were quite numerous, living in the river-bottom and feeding upon seeds and insects. Near Mount Turnbull, also, 1 saw many bevies, though, from the apparently waterless condition of the cañons where they were found, it was not easy to see how they could exist. In the wilderness, they are very shy and wild, but near settlements they seem to lose their suspicion somewhat, and are much more easily approached. They are extremely loath to take wing, and, as they run very swiftly, it is no easy matter to force a bevy to fly, but, when once started, their flight is swift and strong, and usually protracted to a considerable distance. They rarely squat under cover, as the well-known Bob White is wont to do, but usually take the shortest route to the nearest rocky hill, up which they run, and where it is useless to attempt pursuit. From the above characteristics, it will at once be seen that this bird has few qualities to attract the sportsman.

| No. | Sex. | Locality. | Date. | Collector. | Wing. | Tail. | Bill. | Tarsus. |
|-----|------|-----------|-------|------------|-------|-------|-------|---------|
| 777 | ♂ ad. | Gila River, Ariz........ | Sept. 15 | Henshaw . | 4. 33 | 3. 98 | 0. 52 | 1. 23 |
| 891 | ♂ ad. | Camp Grant, Ariz...... | Sept. 27 | .... do .... | 4. 47 | 4. 18 | 0. 50 | 1. 23 |
| 967 | ♂ ad. | Gila River, N. Mex.... | Oct. 25 | .... do .... | 4. 43 | 3. 93 | 0. 45 | 1. 17 |
| 968 | ♀ ad. | ......do .............. | Oct. 25 | .... do .... | 4. 37 | 3. 73 | 0. 45 | 1. 20 |

159. *Callipepla squamata*, (Vigors.)—Scaly Partridge.

Camp Grant was the only locality where this quail was seen. A single small bevy was met with among the bushes on the dry plain. They appeared remarkably unsuspicious, and were very loath to take wing, but, when they did so, flew a long distance, keeping nearly together, and on alighting began to run with remarkable speed, and soon eluded pursuit.

*Description of young.*—Head above grayish-brown, each feather of crest centrally streaked with white; prevailing color of back ashy-brown; tertiaries and interscapular region mottled transversely with rufous and black; wing-coverts centrally streaked and tipped with white; throat ashy-white; under parts generally washed with rufous, and banded, most distinctly on sides, with transverse blackish-brown bars; tail above with indistinct bands, producing the general effect of mottlings of dark-brown and white; bill dark-brown above, lighter below; legs and feet light-brown (in skin).

| No. | Sex. | Locality. | Date. | Collector. | Wing. | Tail. | Bill. | Tarsus. |
|---|---|---|---|---|---|---|---|---|
| 625 | ♂ ad. | Camp Grant, Ariz. .... | Sept. 22 | Henshaw . | 4. 63 | 3. 59 | 0. 48 | 1. 12 |
| 626 | ♀ jun. | ......do .............. | Sept. 22 | .... do .... | 4. 41 | 2. 80 | 0. 45 | 1. 02 |

**160. Cyrtonyx massena, (Lesson).—Massena Partridge.**

This beautiful partridge is quite a common resident in the White Mountains, near Apache, Ariz., where, in summer, it seems to shun the open valleys, and keeps in the open pine-woods, evincing a strong preference for the roughest, rockiest localities, where its stout feet and long, curved, strong claws are admirably adapted to enable it to move with ease. August 10, while riding with a party through a tract of piny woods, a brood of eight or ten young, accompanied by the female, was discovered. The young, though but about a week old, rose up almost from between the feet of the foremost mule, and after flying a few yards dropped down, and in a twinkling were hidden beneath the herbage. At the moment of discovery, the parent bird rose up, and then, tumbling back helplessly to the ground, imitated so successfully the actions of a wounded and disabled bird that, for a moment, I thought she must have been trodden upon by one of the mules. Several of the men, completely deceived, attempted to catch her, when she gradually fluttered off, keeping all the time just beyond the reach of their hands, till she had enticed them a dozen yards away, when she rose and was off like a bullet, much to their amazement. From Apache southward, the species appeared to be quite numerous, always showing its predilection for rocky hills and rough cañons. In the cañons of the Gila River, toward its sources, in New Mexico, in October and November, they were met with frequently, and scarcely a day passed without three or four bevies being flushed. At this season they keep in small bevies, and I do not remember to have ever seen more than ten together, and usually from four to eight. Their tameness and utter want of suspicion is very remarkable, and the more so when contrasted with the wild, timid nature of the Gambel's Partridge, inhabiting the same region. I have ridden so close to a bevy sitting among the rocks, that, leaning down, I could have almost touched them with my hand. When a bevy is flushed, they usually separate, and fly strongly and swiftly in a straight line, dropping down into the first convenient cover. They lie well, requiring to be almost kicked up before taking wing. The species was found in New Mexico as far north as Tulerosa.

*Description of young male.*—Upper parts pale-brown, each feather with a medial, sharply-defined streak of pale-ochraceous, and barred with black across the webs; wing-coverts ashy, with transverse oval or rounded spots of deep black on opposite webs; primaries and secondaries banded transversely with white spots; head grayish-white laterally and beneath; the whole throat unspotted; a dark-brown spot on the auriculars; the region above and below finely streaked with dusky; crown more brownish, spotted with black, and with whitish shaft-streaks; lower parts pale-gray, inclining to plumbeous on middle of breast, each feather with a terminal deltoid spot of white, bordered anteriorly by a narrow bar of black; abdomen tinged with ochraceous; anal region, tibiæ, and coissum velvety-black.

*Chick.*—Head above brownish, with an occipital patch of chestnut-brown; a small black spot behind the eye; crest, of five feathers, just appearing, each feather streaked centrally with white, bordered by blackish-brown; upper parts brown, each feather streaked centrally with white, and with two to three transverse spots of black; under parts dull-white, each feather with transverse spottings of blackish-brown.

| No. | Sex. | Locality. | Date. | Collector. | Wing. | Tail. | Bill. | Tarsus. |
|---|---|---|---|---|---|---|---|---|
| 565 | ♂ ad. | White Mountains, Ariz. | Aug. 12 | Henshaw.. | 4.87 | 2.55 | 0.55 | 1.16 |
| 965 | ♂ jun. | Gila River, N. Mex.. | Oct. 26 | ....do ..... | 4.66 | 2.27 | 0.60 | 1.09 |
| 966 | ♀ jun. | ......do ............ | Oct. 26 | ....do ..... | 4.52 | 2.20 | 0.55 | 1.04 |
| 969 | ♂ jun. | ......do ............ | Oct. 26 | ....do ..... | 4.76 | 2.26 | 0.56 | 1.13 |
| 970 | ♂ jun. | ......do ............ | Oct. 26 | ....do ..... | 4.93 | 2.00 | 0.58 | 1.09 |
| 979 | ♀ ad. | Tulerosa, N. Mex.... | Nov. 15 | ....do ..... | 4.95 | 2.47 | 0.59 | 1.09 |
| 995 | ♀ ad. | Gila River, N. Mex .. | Nov. 15 | ....do ..... | 4.90 | 2.35 | 0.55 | 1.17 |
| 94 | ♀ ad. | Apache, Ariz........ | Sept. 13 | Newberry.. | 4.60 | 2.07 | 0.59 | 1.12 |
|  | ♂ ad. | South of Apache, Ariz. | ......... | Lieutenant Tilman. | 4.75 | ...... | 0.60 | 1.15 |

CHARADRIIDÆ (the Plovers).

161. *Ægialitis vociferus*, (L.)—Killdeer-Plover.

Met with till into November.

SCOLOPACIDÆ (the Snipes).

162. *Gallinago wilsoni*, Temm.—Wilson's Snipe.

In the fall, a few were met with, here and there, in the marshy spots about springs, and even along the open shores of small streams, where they appeared strangely out of place, running along the banks and searching for food much like true Shore-Snipes. A few probably winter.

163. *Macrorhamphus griseus* (Gm.)—Red-breasted Snipe.

Apparently an uncommon visitor in Arizona. A pair were taken at Mimbres by Dr. Newberry, jr. They are in the plumage hitherto known as var. *scolopaceus*, which is now referred by Dr. Coues to the true *griseus*.

| No. | Sex. | Locality. | Date. | Collector. | Wing. | Tail. | Bill. | Tarsus. |
|---|---|---|---|---|---|---|---|---|
| 138 | ♂ ad. | Mimbres, Ariz ........ | Oct. 22 | Newberry. | 5.80 | 2.70 | 2.50 | 1.50 |
| 137 | ♀ ad. | ......do............... | Oct. 22 | .... do .... | 6.00 | 2.80 | 3.00 | 1.60 |

164. *Ereunetes pusillus*, (L.)—Semipalmated Sandpiper.

Small flocks of these little sandpipers were noticed in the marshy spots near the river at Apache. During the migrations, they are doubtless generally distributed, occurring along the water-courses in small numbers.

| No. | Locality. | Date. | Collector. | Wing. | Tail. | Bill. | Tarsus. |
|-----|-----------|-------|-----------|-------|-------|-------|---------|
| 81  | Apache, Ariz ............... | Aug. 29 | Newberry. | 3.74 | 1.78 | 1.05 | 0.88 |
| 644 | ......do .................... | Aug. 28 | Henshaw . | 3.63 | 1.83 | 0.88 | 0.83 |

165. *Actodromus bairdii*, Cs.—Baird's Sandpiper.

Seen along the Zuni River in small numbers, and again at Apache. Nowhere common.

| No. | Sex. | Locality. | Date. | Collector. | Wing. | Tail. | Bill. | Tarsus. |
|-----|------|-----------|-------|-----------|-------|-------|-------|---------|
| 80  | ♀ | Apache, Ariz.......... | Aug. 29 | Newberry. | 4.68 | 2.13 | 0.91 | 0.93 |
| 618 | ♂ jun. | ......do ............... | Aug. 26 | Henshaw . | 4.60 | 2.28 | 0.92 | 0.83 |
| 619 | ♀ jun. | ......do ......--.. | Aug. 26 | .... do .... | 5.00 | 2.33 | 0.90 | 0.87 |
| 619a | ♂ jun. | ......do ............... | Aug. 26 | .... do .... | 4.65 | 2.02 | 0.90 | 0.83 |
| 643 | ♂ | ......do ............... | Aug. 28 | .... do .... | 4.62 | 2.05 | 0.86 | 0.83 |

166. *Actodromus minutilla*, (Vieill.)—Least Sandpiper.

| No. | Sex. | Locality. | Date. | Collector. | Wing. | Tail. | Bill. | Tarsus. |
|-----|------|-----------|-------|-----------|-------|-------|-------|---------|
| 645 | ♀ | Apache, Ariz.......... | Aug. 28 | Henshaw . | 3.36 | 1.77 | 0.73 | 0.73 |

167. *Rhyacophilus solitarius*, (Wils.)—Solitary Sandpiper.

After the latter part of July, quite common, in companies of five or six, along the small water-courses. Several observed near Apache on a small pond in the high pine-woods.

| No. | Sex. | Locality. | Date. | Collector. | Wing. | Tail. | Bill. | Tarsus. |
|-----|------|-----------|-------|-----------|-------|-------|-------|---------|
| 84  | ♀ | Apache, Ariz.......... | July 29 | Newberry. | 5.29 | 2.47 | 1.24 | 1.24 |
| 509 | ♂ | Cave Spring, Ariz..... | Aug. 1 | Henshaw . | 5.50 | 2.66 | 1.20 | 1.23 |
| 510 | ♂ | ......do ............... | Aug. 1 | .... do .... | 5.22 | 2.32 | 1.25 | 1.24 |
| 541 | ♀ | Near Apache, Ariz .... | Aug. 9 | .... do .... | 5.25 | 2.27 | 1.20 | 1.20 |
| 546 | ...... | ......do ............... | Aug. 10 | .... do .... | 5.12 | 2.33 | 1.29 | 1.26 |
| 548 | ♀ | ......do ............... | Aug. 10 | .... do .... | 5.23 | 2.35 | 1.17 | 1.20 |
| 650 | ...... | ......do ............... | Aug. 28 | .... do .... | 5.35 | 2.55 | 1.20 | 1.16 |

168. *Tringoides macularius*, (L.)—Spotted Sandpiper.

Frequents the water-courses.

RECURVIROSTRIDÆ (the Stilts).

169. *Recurvirostra americana*, Gm.—American Avocet.

Common on the Colorado Chiquito in August.  (Newberry.)

10 o s

, GRUIDÆ (the Cranes).

170. *Grus canadensis*, L.—Brown Crane; Sandhill-Crane.
A few seen in the valleys along the streams. It is fond of frequenting the old stubble-fields in the vicinity of settlements.

TANTALIDÆ (the Ibises).

171. *Ibis guarauna*, (L.)—White-faced Ibis.
A specimen was brought to me at Apache by an Indian, which he had shot with his rifle. Not seen elsewhere.

ARDEIDÆ (the Herons).

172. *Ardea herodias*, L.—Great Blue Heron.
Of common occurrence along the streams.

173. *Herodias alba* (L.), var. *egretta*, (Gm.)—Great White Egret.
A single individual was seen on a small creek at Camp Grant, but was so wary that all attempts to capture it proved unavailing. One seen also on the San Pedro River.

174. *Butorides virescens*, (L.)—Green Heron.
One seen at Camp Grant.

175. *Nyctiardea grisea* (L.), var. *nævia*, Bodd.—Night-Heron.
Several seen on the Colorado Chiquito by Dr. Newberry, jr.

RALLIDÆ (the Rails).

176. *Rallus virginianus*, L.—Virginia Rail.
Two were put up from a bed of reeds along the Apache River.

177. *Porzana carolina*, (L.)—Carolina Rail.
Two seen in same locality. (Dr. Newberry, jr.)

178. *Fulica americana*, Gm.—Coot.
Abundant on the San Pedro River, Arizona, the first of October. Many also seen in a beaver-pond near Tulerosa, N. Mex.

ANATIDÆ (the Ducks).

From the dry and generally waterless nature of much of the country traversed by the survey, comparatively few of the waders and swimmers were seen. Occasionally, however, in places suited to their habits, as the sloughs along the San Pedro River, they were seen to be abundant, and a few were seen, from time to time, on the small streams. Without doubt, the greater proportion of the ducks found in Western Interior occur during the migrations in Arizona and New Mexico.

179. *Dafila acuta*, (L.)
Abundant on the San Pedro.

**180.** *Anas boschas,* L.

Frequent.

**181.** *Nettion carolinensis,* (Gm.)—Green-winged Teal.

**182.** *Querquedula discors,* (L.)—Blue-winged Teal.

These two species are perhaps the most common and generally-distributed of the family. Nearly every little stream and pond-hole containing sufficient water to float them will be found occupied by a flock of either of the two species, or often both mingled together.

**183.** *Spatula clypeata,* (L.)—Shoveler.

Numerous.

**184.** *Erismatura rubida,* (Wils.)—Ruddy Duck.

One shot at the Old Crater, south of Zuni, N. Mex.

LARIDÆ (the Gulls and Terns).

**185.** *Sterna hirundo,* L.—Common Tern.

A single specimen was shot on the San Pedro River, Arizona, in September. The river at this point was but a small stream, perhaps twenty feet across, and the bird was flying slowly up this, closely scanning the water for fish. The most western point at which this bird has hitherto been known to occur is Wisconsin.

| No. | Sex. | Locality. | Date. | Collector. | Wing. | Tail. | Bill. | Tarsus. |
|---|---|---|---|---|---|---|---|---|
| 897 | ♀ jun. | San Pedro River, Ariz. | Sept. 3 | Henshaw. | 9.75 | 4.83 | 1.10 | 0.70 |

LIST OF EGGS COLLECTED IN 1873.

| No. | Name. | Locality. | Date. | Collector. |
|---|---|---|---|---|
| 155 | Turdus migratorius | Garland, Colo | June — | H. W. Henshaw, 2 eggs. |
| 149 | Turdus, var. audubonii | ......do | June 7 | Do. |
| 151 | Turdus fuscescens | ......do | June 7 | H. W. Henshaw, 4 eggs. |
| 255 | Oreoscoptes montanus | Alkali lakes, Colo | June 22 | Do. |
| 34 | ......do | Wingate | July 14 | Dr. Newberry, 2 eggs. |
| 254 | Galeoscoptes carolinensis | Garland, Colo | July 14 | H. W. Henshaw, 4 eggs. |
| 203 | Dendroica æstiva | Alkali lakes, Colo | June 22 | H. W. Henshaw, 5 eggs. |
| 337 | Poœcætes var. confinis | Garland, Colo | .......... | H. W. Henshaw, 4 eggs. |
| 82 | ......do | South Park, Colo | July 1 | Dr. Rothrock, 2 eggs. |
| 398 | Pipilo chlorurus | Garland, Colo | June 10 | H. W. Henshaw, 6 eggs. |
| 271 | Troglodytes, var. parkmanni | ......do | July — | H. W. Henshaw, 2 eggs. |
| 268 | Telmatodytes palustris | Alkali lakes, Colo | .......... | H. W. Henshaw, 1 egg. |
| B 5 | Chondestes grammaca | Colorado | July — | Dr. Rothrock, 4 eggs. |
| 418 | Scol. cyanocephalus | Garland, Colo | June–July | H. W. Henshaw, 30 eggs. |
| 404 | Xan. icterocephalus | ......do | June 23 | H. W. Henshaw, 2 eggs. |
| 136 | Sayornis sayus | ......do | June 19 | H. W. Henshaw, 3 eggs. |
| 226 | Petroch. luuifrons | ......do | .......... | H. W. Henshaw, 16 eggs. |
| 104 | Sel. platycercus | ......do | June 14–19 | H. W. Henshaw, 4 eggs. |
| 451 | Zenaidura carolinensis | Denver, Colo | May 5–14 | H. W. Henshaw, 16 eggs. |
| 459 | Canace obscura | Rio Grande, Colo | June 16 | H. W. Henshaw, 1 egg. |
| 518 | Himantopus nigricollis | Alkali lakes, Colo | June 22 | H. W. Henshaw, 4 eggs. |
| 517 | Recurvirostra americana | ......do | June 22 | Do. |
| 519 | Nettion carolinensis | ......do | June 23 | H. W. Henshaw, 10 eggs. |
| 409 | Podiceps auritus, var. californicus | ......do | June 23 | H. W. Henshaw, 23 eggs. |
| 559 | Fulica americana | ......do | June 23 | H. W. Henshaw, 35 eggs. |

LIST OF NESTS COLLECTED IN 1873.

| No. | Name. | Locality. | Date. | | Collector. |
|---|---|---|---|---|---|
| 369 | Turdus, var. audubonii ........ | Garland, Colo........... | June | 7 | H. W. Henshaw. |
| 151 | Turdus fuscescens ............. | ......do ................. | June | 19 | H. W. Henshaw (2 nests built together). |
| 368 | Dendroica audubonii ............ | ......do ................. | June | 8 | H. W. Henshaw. |
| 396 | Dendroica æstiva................. | ......do ................. | June | 22 | Do. |
| 381 | ......do ...................... | ......do ................. | June | 20 | Do. |
| 82 | Pooecetes, var. confinis .. ........ | South Park, Colo........ | July | 1 | Dr. Rothrock. |
| 420 | ......do ...................... | Garland, Colo ............ | July | 19 | H. W. Henshaw. |
| B5 | Chondestes grammaca .......... | Colorado................. | July | — | Dr. Rothrock. |
| 397 | Pipilo —— ................. | Alkali lakes. Colo ...... | ........... | | H. W. Henshaw. |
| 365 | Pipilo chlorurus ............... | Rio Grande River, Colo.. | June | 15 | Do. |
| 406 | Agelaius phœniceus...... ...... | Garland, Colo............ | June | 21 | Do. |
| 156 | Scolecophagus cyanocephalus... | ......do ................. | May | 27 | Do. |
| 155 | ......do ...................... | ......do ................. | May | 27 | Do. |
| 71 | Small Flycatcher .............. | South Park, Colo ....... | June | 30 | Dr. Rothrock. |
| 30 | Tyrannus verticalis.............. | Denver, Colo ............ | June | 10 | Do. |
| 31 | ......do ...................... | ......do ................. | June | 10 | Do. |
| 351 | Selasphorus platycercus ... ..... | Rio Grande, Colo ....... | June | 14 | H. W. Henshaw. |
| 365 | ......do ...................... | Garland, Colo ........... | June | 19 | Do. |
| 31 | Zenaidura carolinensis.......... | Denver, Colo............ | June | 10 | Dr. Rothrock. |
| C4 | ......do ...................... | ......do ................. | June | — | Do. |

LIST OF STERNA COLLECTED IN 1873.

| No. | Name. | Sex. | Locality. | Date. | Collector. |
|---|---|---|---|---|---|
| 390 | Recurvirostra americana......... | ♂ ad. | Alkali lakes, Colo ...... | June 21 | Henshaw. |
| 391 | ......do ...................... | ♀ ad. | ......do ................. | June 21 | Do. |
| 389 | ......do ...................... | ♂ ad. | ......do ................. | June 21 | Do. |
| 398 | Himantopus nigricollis........... | ♀ | ......do ................. | June 22 | Do. |
| 417 | Querquedula cyanoptera.......... | ♂ ad. | ......do ................. | June 23 | Do. |
| 306 | Ægialitis montanus .............. | ♂ ad. | Garland, Colo ........... | June 10 | Do. |
| 307 | ......do ...................... | ♂ | ......do ................. | June 10 | Do. |
| 309 | Picicorvus columbianus.......... | ♂ ad. | Rio Grande, Colo....... | June 10 | Do. |
| 310 | ......do ...................... | ♂ | | | Do. |
| 190 | ......do ...................... | ♂ | Garland, Colo .......... | May 29 | Do. |
| 616 | Falco polyagrus ........ ........ | ♀ jun. | Apache, Ariz ......... | Aug. 26 | Do. |
| 64 | ......do ...................... | ♂ | Denver, Colo............ | May 12 | Do. |
| 35 | Nisus fuscus.................... | | ......do ................. | May 9 | Do. |
| 277 | Myindestes townsendii........... | ♂ | Garland, Colo........... | June 6 | Do. |
| 503 | Calamospiza bicolor............. | ♂ | Pescao, N. Mex ......... | July 25 | Do. |
| 414 | Podiceps, var. californicus....... | ♂ ad. | Alkali lakes, Colo ...... | June 23 | Do. |
| 612 | Antro. nuttalli...... .......... | ♀ jun. | Apache, Ariz ......... | Aug. 25 | Do. |
| 90 | ......do ...................... | ♂ | Denver, Colo ........... | May 15 | Do. |
| 220 | Sphy. thyroideus ............... | ♀ | Garland, Colo........... | June 3 | Do. |
| 219 | ......do ...................... | ♂ ad. | ......do ................. | June 3 | Do. |
| 235 | ......do ...................... | ♂ ad. | ......do ................. | June 3 | Do. |
| 298 | ......do ...................... | ♂ | ......do ................. | May 20 | Do. |
| 234 | ......do ...................... | ♂ ad. | ......do ................. | June 4 | Do. |
| 221 | Picoides dorsalis ............... | | ......do ................. | June 3 | Do. |
| 202 | Perisoreus var. capitalis ......... | ♂ | ......do ................. | May 30 | Do. |
| 224 | ......do .......... ........... | ♀ | ......do ................. | June 3 | Do. |
| 223 | ......do ...................... | ♂ jun. | ......do ................. | May 30 | Do. |

www.ingramcontent.com/pod-product-compliance
Lightning Source LLC
Chambersburg PA
CBHW021814190326

41518CB00007B/593